# THE GEOLOGICAL HISTORY of NEW YORK STATE

Taughannock Falls (height 215 feet) which ranks as the highest true waterfall in New York State. Located near Trumansburg, Tompkins county. The rocks are Devonic sandstones and sandy shales and the postglacial gorge just below the falls is nearly 400 feet deep.

Photo loaned by H. R. Head, Ithaca, N. Y.

William J. Miller

# THE GEOLOGICAL HISTORY of NEW YORK STATE

KENNIKAT PRESS
Port Washington, N. Y./London

KENNIKAT PRESS SCHOLARLY REPRINTS
Dr. Ralph Adams Brown, Senior Editor

Series on
MAN AND HIS ENVIRONMENT
Under the General Editorial Supervision of
Dr. Roger C. Heppell
*Professor of Geography, State University of New York*

THE GEOLOGICAL HISTORY OF NEW YORK STATE

First published in 1914 by the University of the
State of New York as Museum Bulletin 168
Reissued in 1970 by Kennikat Press
Library of Congress Catalog Card No: 73-113291
ISBN 0-8046-1326-5

Manufactured by Taylor Publishing Company Dallas, Texas

KENNIKAT SERIES ON MAN AND HIS ENVIRONMENT

# CONTENTS

# THE GEOLOGICAL HISTORY OF NEW YORK STATE

BY

WILLIAM J. MILLER Ph.D.

## PREFACE

The researches and truths of any modern science if they are properly to fulfil their mission, should be brought within the reach of laymen. In this bulletin the purpose is to present in a simple, readable form, an outline of the wonderful story of the physical development of New York State. No knowledge of physiography or geology is presupposed. Any person who possessed of intelligence and a willingness to learn is fully prepared to read these pages.

When the reader has gained a fair understanding of the principles here set forth, he will be much better prepared to use intelligently the publications of the New York State Museum which deal with the geology and geography of many portions of the State. In short, this volume may be considered as a " first book " for all who are interested in the physical features of our State and it is believed that teachers and older pupils in geography and physical geography may receive helpful suggestions from it.

It must be clearly understood that scarcely more than a sketch of such a large subject can be given in so brief a space. Local

details can seldom be brought in except for illustration of certain important points, and of necessity many questions will occur to readers interested in the natural features of their home regions which are not directly answered. It is hoped, however, that most of the important and striking geographic features in all parts of the State are explained, and that many local details will find ready explanation by the application of the principles set forth.

Emphasis is here placed upon the genesis of geographic forms. It is one thing merely to state a geographic fact, such as the location of a mountain or lake or valley, but it is a far different thing to explain how the mountain or lake or valley came to be there. Every geographic form has a history, and if we fail to appreciate that history we lose the most interesting and valuable part of our geographic training. Geographic facts, like all others, are more easily understood and remembered when the reasons for their existence are given, yet it must be admitted that the teaching of such rational geography is still in its infancy in the schools of this State.

The use of a certain number of scientific terms is unavoidable in practice, but common terms only are employed and in every case these are carefully explained when first used in the text. Particular attention is directed to the photographs, maps and diagrams, all of which have been carefully selected or made for the express purpose of illustrating this text. Except for some quotations, references to original papers have been omitted, but at the end of the volume a list of the more important books and papers of general interest is given, and anyone desiring to broaden into wider fields or greater details can readily do so with the aid of those references.

I have used many personal observations made during travels into almost every county of the State, but obviously the book could never have been written were it not for scores of devoted men of science who, during the last hundred years, have zealously labored to unravel the natural history of our great Commonwealth. I gratefully acknowledge my indebtedness to them all.

I am under particular obligation to Dr J. M. Clarke, our able and efficient State Geologist and Director of Science, for his kindness in critically reading the manuscript and making important corrections and suggestions.

W. J. M.

*Hamilton College, Clinton, N. Y.*

## Chapter 1

# INTRODUCTION

### GENERAL PRINCIPLES AND REFERENCE TABLES

Few states present a more wonderful variety of physical features or afford a more excellent opportunity to those interested in the study and teaching of geography or geology than does New York. Here are rock formations of all the more important types; all the leading types of mountains (Adirondacks, Catskills, and Taconics) except actual volcanoes, and even true lavas occur in the Adirondacks and in the Palisades of the Hudson; hundreds of lakes of various shapes and kinds; shore outlines ranging from the great sand bars and beaches of Long Island to wave-worn cliffs along the shores of Lake Erie and Lake Ontario; typical prairie plains like that south of Lake Ontario; a great plateau in the southwestern region; valleys and gorges of varied origin; rivers of all types and often with remarkable histories; a striking display of relief features; and extensive and varied deposits of glacial origin. Accordingly, it is not an exaggeration to say that examples of nearly all the most important physical features of the earth are represented within the borders of this State.

As the observer looks out over the State he sees this great variety of physical features and, unless he has given some thought to the subject, is very likely to regard these as practically unchangeable, and that they are now essentially as they were in the beginning of the earth's history. Some of the fundamental ideas taught in this book are that the physical features of the State, as we behold them today, represent but a single phase of a very long continued history; that significant changes are now going on all around us; and that we are able to interpret the geography of the present only by an understanding of its changes in the past.

*Geology* is concerned with the evolution of the earth and of its inhabitants, as revealed in the rocks. This science is very broad in its scope and treats of the processes by which the earth has been, and is now being, changed; the structure of the earth; the stages through which it has passed, and the development of the organisms which have lived upon it.

*Geography* deals with the distribution of the earth's physical features, in their relation to each other, to the life of sea and land and human life and culture.

Geography is the outward and present expression of geological

effects. The terms geography and geology are thus here used in the sense that the latter includes the former, as the cause includes the effect. *Paleogeography* has reference to the geography of the past epochs in the history of the earth.

*Physiography,* or physical geography, deals with the configuration (relief) of the earth's surface and how it was produced.

As a result of the work of many able students of earth science during the past hundred years, it is now well established that our planet has a clearly recorded history of many millions of years, and that during the lapse of those eons revolutionary changes in geography have occurred; that there has also been from an early stage of the earth's history a vast succession of living beings which have gradually passed from simple into more complex forms and have, in some particulars, reached their highest expression in the organisms of the present time. The geographic changes and the organisms of the ages gone by have left us no abundant evidence of their character and the study of the rock formations has shown that within them we have a fairly complete record of the earth's history.

In the time of Alexander von Humboldt, less than one hundred years ago, the keen student of natural phenomena could carry in his own mind most of what was definitely known of earth history. Today, because of the tremendous growth of the science, it would be a presumption for any man to claim that he knows all of what has been learned about the geological history of even the single State of New York. While it is true that much yet remains to be learned of this old earth, it is a real source of wonderment that man, through the exercise of his highest faculty, has come to know so much about it.

All the rocks of the earth's crust may be divided into three great classes: *igneous, sedimentary,* and *metamorphic.*

Igneous rocks comprise all those which have ever been in a molten condition, and of these we have the volcanic rocks (for example, lavas) which have cooled at or near the surface; plutonic rocks (e. g., granites) which have cooled in great masses at considerable depths below the surface; and the dike rocks, which when molten have been forced into fissures of the earth's crust and there cooled.

Sedimentary rocks comprise all those which have been deposited under water (except for some wind-blown deposits) and are nearly always arranged in layers (stratified). These rocks may be of mechanical origin, such as clay or mud which hardens to shale; sand, which consolidates to sandstone; and gravel, which when cemented becomes conglomerate. Or they may be of organic origin,

such as limestone which is formed by the accumulation of calcareous shells; flint and chert, which are accumulations of silicious shells; coal, which is formed by the accumulation of partly decayed vegetable matter. Or, finally, they may be formed by chemical precipitation, as beds of salt, gypsum, bog iron ore, etc.

Metamorphic rocks comprise both sedimentary and igneous masses which have been greatly changed from their original condition. Thus, under conditions of great pressure and heat, with superheated moisture, sedimentary rocks may become crystalline, as when shale is changed to schist, sandstone to quartzite, or limestone to marble; or an igneous rock may take on a banded structure, due to a rearrangement of its component minerals, and thus become a gneiss.

To the modern student of earth science, the old notion of a "terra firma" is outworn. That idea of a solid, immovable earth could never have emanated from the inhabitants of an earthquake country. In the recent San Francisco earthquake, along a line of several hundred miles, one portion of the Coast Range mountains slipped from two to twenty feet past the other. In Alaska, in 1899, a portion of the coast was bodily elevated forty-seven feet. In Japan in 1891, for a distance of forty miles along a rift in the earth's crust, there was a sudden movement of from two to twenty feet. These are merely striking instances of many of the sudden earth movements of recent years. Hundreds of earthquakes occur yearly in the islands of Japan alone, and it is probably true that the earth is shaking all the time.

There are still other movements which are taking place more slowly and quietly, but which are more significant for our interpretation of the profound geographic changes which have occurred during the millions of years of known earth history. Thus the coast of Norway is rising while that of northern France is sinking. Distinct beaches at different elevations far above the ocean level on the western slope of the southern Andes testify to important changes of level in comparatively recent time. A fine illustration of notable sinking of the land is proved by the drowned character of the lower Hudson valley, and by the fact that the old Hudson channel has been definitely traced, as a distinct trench in the ocean bottom, for one hundred miles eastward from Sandy Hook. That this same region has still more recently been partially re-elevated is indicated by the presence of very young stratified beds of clay and sand which are now raised from seventy to three hundred feet above the river, the elevation increasing northward toward Albany. Actual surveys show that, in the Great Lakes region, a differential

movement of the land is now in progress, the elevation being greater toward the north.

In the succeeding pages evidence will be offered to show that most of New York State has been repeatedly covered by ocean water. It will also be established that where such mountain ranges as the Appalachians, Alps or Himalayas now exist was formerly ocean bottom upon which layers of sediment were being spread out. Those layers of sediments have been bent, crumpled, folded, and greatly elevated above sea level. Thus it is literally true that the great typical mountain ranges of the earth have been born out of the ocean.

Among other important processes of nature which have long been active in modifying the earth, are those of *weathering* and *erosion.* Weathering is brought about by the various atmospheric agencies such as moisture, oxygen, carbonic and other acids, together with changes of temperature, and the result is to cause all rock masses to disintegrate or decay. In this way most soils are produced, and were it not for the process of erosion, soils would be much deeper and more widespread than they now are. Weathering prepares the material which is carried away by the streams, and this transported material is deposited either along the flood plains of the lower stream courses or on the bottom of the lake or ocean into which the streams flow. Every stream, at time of flood, is heavily charged with mud or even coarser sediment which has been derived from the wash of the land of its drainage basin. The very presence of the sediment in the streams proves that the land is being lowered and although, on first thought, it may be supposed that no really great change could be accomplished by this means, nevertheless we must remember that nature has practically infinite time at her disposal so that slowly but surely vast geographic changes are wrought and, perchance, a tremendous canyon like that of the Colorado in Arizona will be carved out by weathering and erosion. The general tendency is for all land masses to wear down to or near sea level and, were it not for renewed uplifts, all land, even including mountain ranges, would long ago have been worn down to near sea level, that is to the condition of *peneplain* (almost a plain). The former lofty Appalachians were thus worn down to the condition of a peneplain which has since been somewhat rejuvenated by elevation. Accordingly, that familiar expression "the everlasting hills" is much more exact when made to read "the everlastingly changing hills."

Still another important process by which the physical features of the earth have often been changed is through *vulcanism,* or

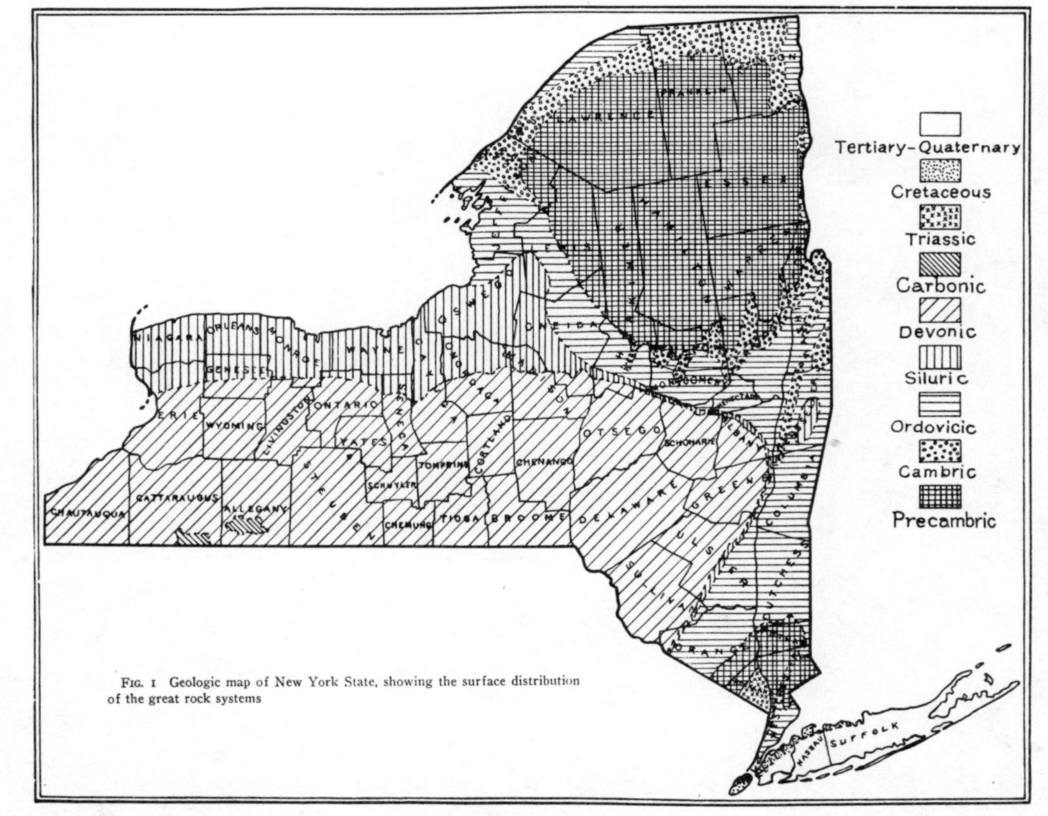


Fig. 1 Geologic map of New York State, showing the surface distribution of the great rock systems

igneous activity in general. By this means materials are brought up from within the earth to or near its surface. Thus an active volcano violently ejects rock fragments, dust etc. or more quietly pours out molten rock, while in many cases great masses of molten rocks have been forced upward into the crust of the earth without reaching the surface and hence have slowly cooled at greater or lesser depths below the surface. Such volcanic rocks have become exposed to view only by subsequent erosion of the region.

In order to understand the physical history of our State it is necessary to know that significant changes, like those above described, have long been, and now are taking place. In tracing this history we shall see how all these natural processes have operated to bring the State into its present condition. It is also necessary to understand that the known history of the earth has been carefully divided into great eras, and into lesser periods and epochs, and that these constitute what is called the geologic time scale. This time scale is important to the reader because the principal events in the history of the State will be taken up, so far as they are recorded, in regular order according to that scale. In the first table the names of eras and periods are mostly of world-wide usage, while the names of subdivisions (epochs) of the periods are much more local in usage and, in the second table, only those are given which apply to New York State.

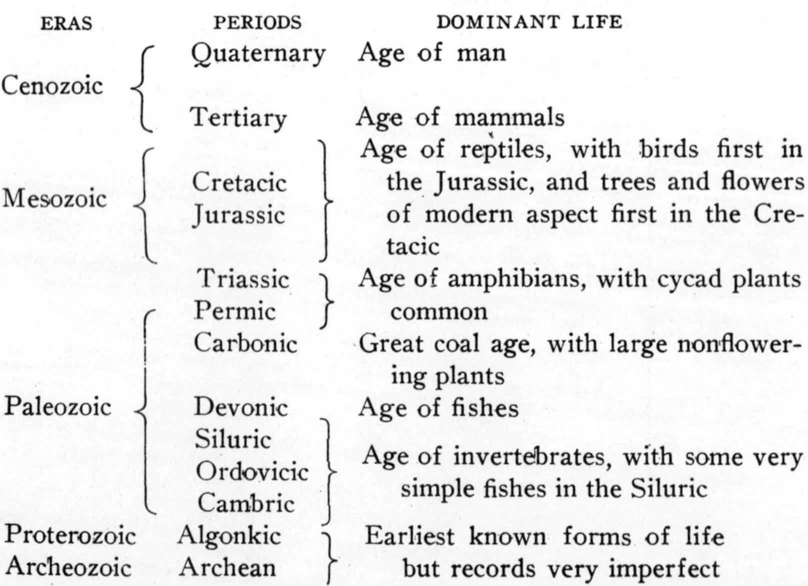

Geologic Time Scale

| ERAS | PERIODS | DOMINANT LIFE |
|---|---|---|
| Cenozoic | Quaternary | Age of man |
| | Tertiary | Age of mammals |
| Mesozoic | Cretacic<br>Jurassic | Age of reptiles, with birds first in the Jurassic, and trees and flowers of modern aspect first in the Cretacic |
| | Triassic | Age of amphibians, with cycad plants common |
| Paleozoic | Permic | |
| | Carbonic | Great coal age, with large nonflowering plants |
| | Devonic | Age of fishes |
| | Siluric<br>Ordovicic<br>Cambric | Age of invertebrates, with some very simple fishes in the Siluric |
| Proterozoic<br>Archeozoic | Algonkic<br>Archean | Earliest known forms of life but records very imperfect |

## PERIODS, EPOCHS, AND ROCK FORMATIONS IN NEW YORK STATE

| Period | Epochs and formations | Where found |
| --- | --- | --- |
| Quaternary | Recent clay beds, alluvium etc.<br>Pleistocene. Glacial clay, sand, gravel, boulders etc. | Surface deposits very common over the State |
| Tertiary | Pliocene<br>Miocene | Sands, clays, and gravels on Long Island |
| Cretacic | Monmouth<br>Matawan<br>Magothy<br>Raritan | Clay and sand on Long and Staten Islands |
| Jurassic | Absent (probably) | |
| Triassic | Newark sandstone and Palisade lava | Rockland county |
| Permic | Absent | |
| Carbonic | Olean conglomerate<br>Cattaraugus shale | Allegany and Cattaraugus counties |
| Devonic | Catskill and Chemung sandstones | Catskills and southwestern New York |
| | Portage shale and sandstones | South-central and western New York |
| | Genesee shale<br>Tully limestone | East-central to western New York |
| | Hamilton shale<br>Marcellus shale<br>Onondaga limestone<br>Oriskany sandstone<br>Helderberg limestone | East-central to western New York and eastern base of Catskill mountains |
| Siluric | Manlius limestone<br>Rondout waterlime<br>Cobleskill limestone<br>Salina shales, salt, waterlime and Shawangunk conglomerate | Eastern to western New York except Shawangunk in eastern New York only |
| | Lockport and Guelph dolomites<br>Rochester shale | Central to western New York |
| | Clinton shale, sandstone, limestone and iron ore | Central to western New York |
| | Medina and Oneida sandstone and conglomerate | Central to western New York |

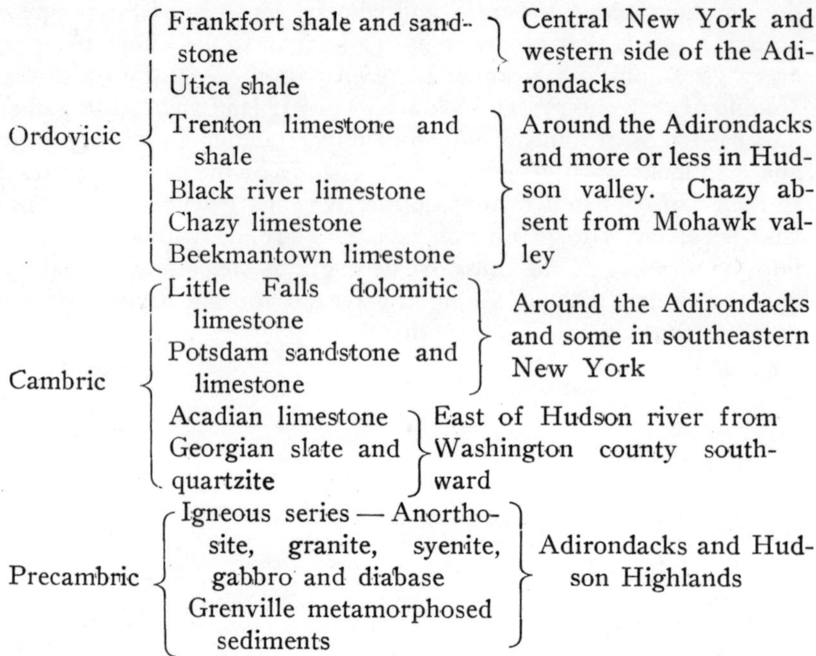

| System | Formations | Distribution |
|---|---|---|
| Ordovicic | Frankfort shale and sandstone<br>Utica shale | Central New York and western side of the Adirondacks |
| | Trenton limestone and shale<br>Black river limestone<br>Chazy limestone<br>Beekmantown limestone | Around the Adirondacks and more or less in Hudson valley. Chazy absent from Mohawk valley |
| Cambric | Little Falls dolomitic limestone<br>Potsdam sandstone and limestone | Around the Adirondacks and some in southeastern New York |
| | Acadian limestone<br>Georgian slate and quartzite | East of Hudson river from Washington county southward |
| Precambric | Igneous series — Anorthosite, granite, syenite, gabbro and diabase<br>Grenville metamorphosed sediments | Adirondacks and Hudson Highlands |

Throughout this book the purpose is not merely to describe the physical features of the State, but rather constantly to emphasize the history or evolution of those features. The idea which I would now convey to the reader has been admirably expressed by Professor Davis in his " The Physical Geography of Southern New England ": " Geography still retains too much of its old-fashioned, irrational methods: it has not kept pace with the advance made by geology. In spite of what the geologist has learned about the evolution of geographical forms, the geographer still too generally treats them empirically, and thus loses acquaintance with one of the most interesting phases of his subject. . . . It is often maintained that a devoted study of the facts themselves, without regard to their meaning or development, will suffice to place them clearly enough before the mind; but this view is contradicted both by general experience in many subjects where rational explanation has replaced empirical generalization, and by the special experience of geography as well. Left to itself as an empirical study, in which the development of land forms was hardly allowed to enter, it has languished for many years, until it became a subject for continual complaint. . . . Today it is only by those who fail to see the direc-

tion of geographical progress, and who are ignorant of the progress already gained, that objection is made against the effort to bring every geographical fact under the explanation of natural processes. No one of active mind can look across our upland and fail to gather increased pleasure and profit from understanding its history. No one who looks upon geography as the study of the earth in relation to man can contemplate the contrast between glaciated New England [or New York] and nonglaciated Carolina without inquiring into the meaning of the contrast: he might as well study the Sahara and the Sudan without asking the reason for the dryness of the one and the moisture of the other."

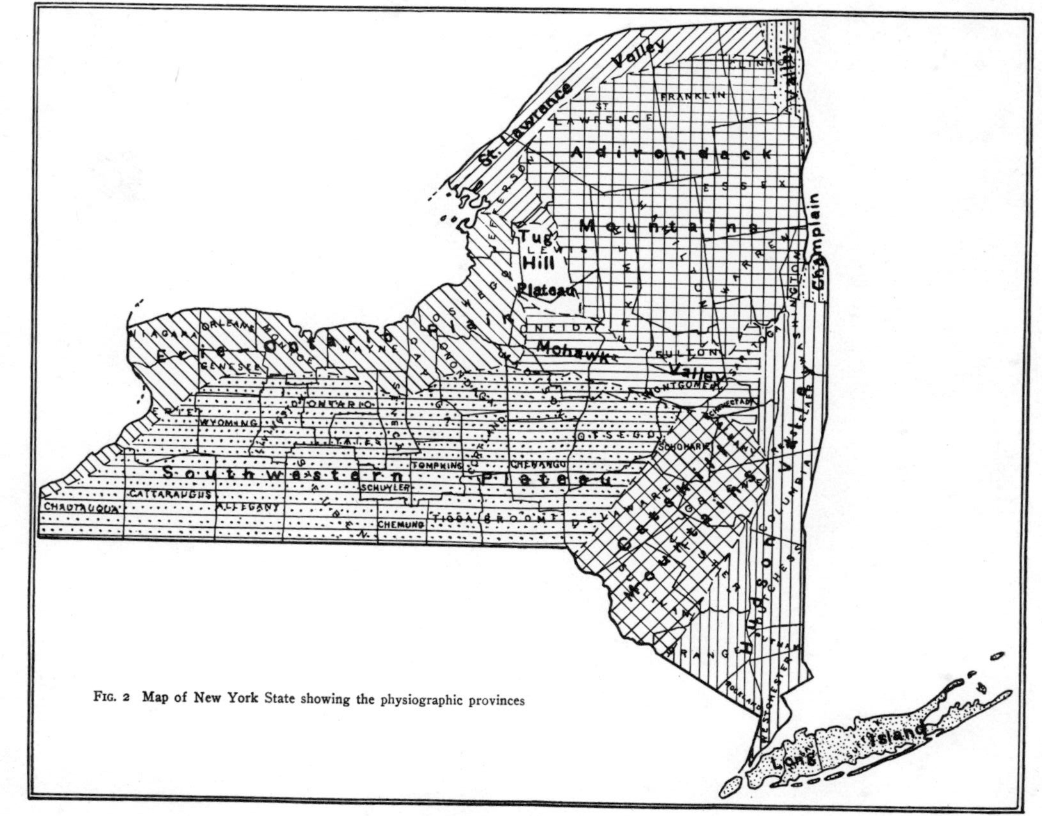


Fig. 2 Map of New York State showing the physiographic provinces

## Chapter 2

# PHYSIOGRAPHIC PROVINCES, STRUCTURE AND DRAINAGE

### GENERAL STATEMENT

The area of this State is 49,170 square miles, including 1550 square miles of water. The range in altitude is from sea level to over 5000 feet, while the average elevation is about 900 feet. Mt Marcy (altitude 5344 feet) in Essex county is the highest mountain in the State.

For the sake of convenience in discussing the general physiography and structure, the writer has divided the State into certain well-defined physiographic provinces as shown on the accompanying map. Lest the sharp boundary lines convey a wrong impression, it should be stated that the provinces are, in reality, seldom sharply separated from each other (see figure 2).

### ADIRONDACK MOUNTAIN PROVINCE

The Adirondack mountain province comprises fully one-fourth the area of the State and consists of a great, nearly circular mass of metamorphic and igneous rocks of very great age, that is, Prepaleozoic. This large mass of crystalline rocks is completely surrounded by the practically unaltered Cambric and Ordovicic rocks. The whole province is typically mountainous and heavily wooded, often being truly wilderness in character with very few roads or settlements other than summer resorts. Except along the immediate borders, the elevations range from 1000 to over 5000 feet. The greatest axis of elevation extends from southern Hamilton county (2000 feet) northeasterly well into Essex county where the highest mountains are grouped around Mt Marcy, and where the mountains commonly attain altitudes of from 3000 to 5000 feet (see plate 1). In the eastern and southeastern portions there is a well-defined tendency in the mountain masses to be arranged in long, nearly parallel ridges or " ranges " whose general trend is north-northeast to south-southwest. This structural feature is due to numerous faults or fractures in the earth's crust and will be explained on a later page. In the northern and western portions the mountains are very irregularly arranged. Viewed as a whole there are no high, sharp-topped peaks which stand out prominently above the general mountain level, and the flowing or rounded outline of topography is by far the most common (see plates 2 and 3). The very ancient Grenville rocks occur throughout the

region, and great masses of igneous rocks have been forced through these. All these rocks have been subjected to tremendous earth pressure which has folded and thoroughly metamorphosed them.

## SOUTHWESTERN PLATEAU PROVINCE

This, the largest clearly defined physiographic province, occupies nearly one-third of the area of the State. The rocks are all unaltered sediments of Devonic age, except a few small patches of Carbonic rocks in the southwest, and consist of shales, sandstones, and conglomerates. These formations exist as vast sheets or layers piled one upon another, with an aggregate thickness of several thousand feet (see figures 3 and 5). In marked contrast to the Adirondack province, the rock masses of this southwestern plateau are practically devoid of displacements, the only disturbance being a slight tilt (30 to 50 feet a mile) of all the strata to the south or southwest, associated with low northeast-southwest undulation.

Although this plateau is pretty well trenched or dissected by streams, it is nevertheless not a mountainous country, there being no high ranges or peaks standing out prominently. The elevation of the province varies from 500 to 600 feet on the northern side to over 2000 feet on the eastern and western sides. A notable feature is the distinct sagging of the plateau toward the middle portion. This sagged or depressed portion is occupied by the Finger Lakes, especially that portion filled by the south ends of Cayuga and Seneca lakes and the valleys which enter them from the south in the region of Chemung and Tioga counties. It is possible, by traveling along Seneca lake and thence southward to Elmira and the Chemung river, to pass entirely across the plateau province from north to south without attaining an altitude of much over 900 feet, which is on the divide between Watkins and Elmira.

Physiographically, the Plateau province is really but the northernmost extension of the great plateau which lies along the western base of the Appalachian mountains. On the east the province is bounded by the Catskill mountains which are in no sense sharply separated from the plateau itself. On the west and north the province is bounded by the Erie-Ontario plain and Mohawk valley provinces. The northern limit is pretty clearly marked by what is known as the "Helderberg escarpment" of Devonic limestone. This limestone, being of considerable thickness and more resistant than the neighboring formations, has generally stood out boldly against erosion, thus causing an abrupt change in relief. The escarpment is particularly prominent along the boundary of the

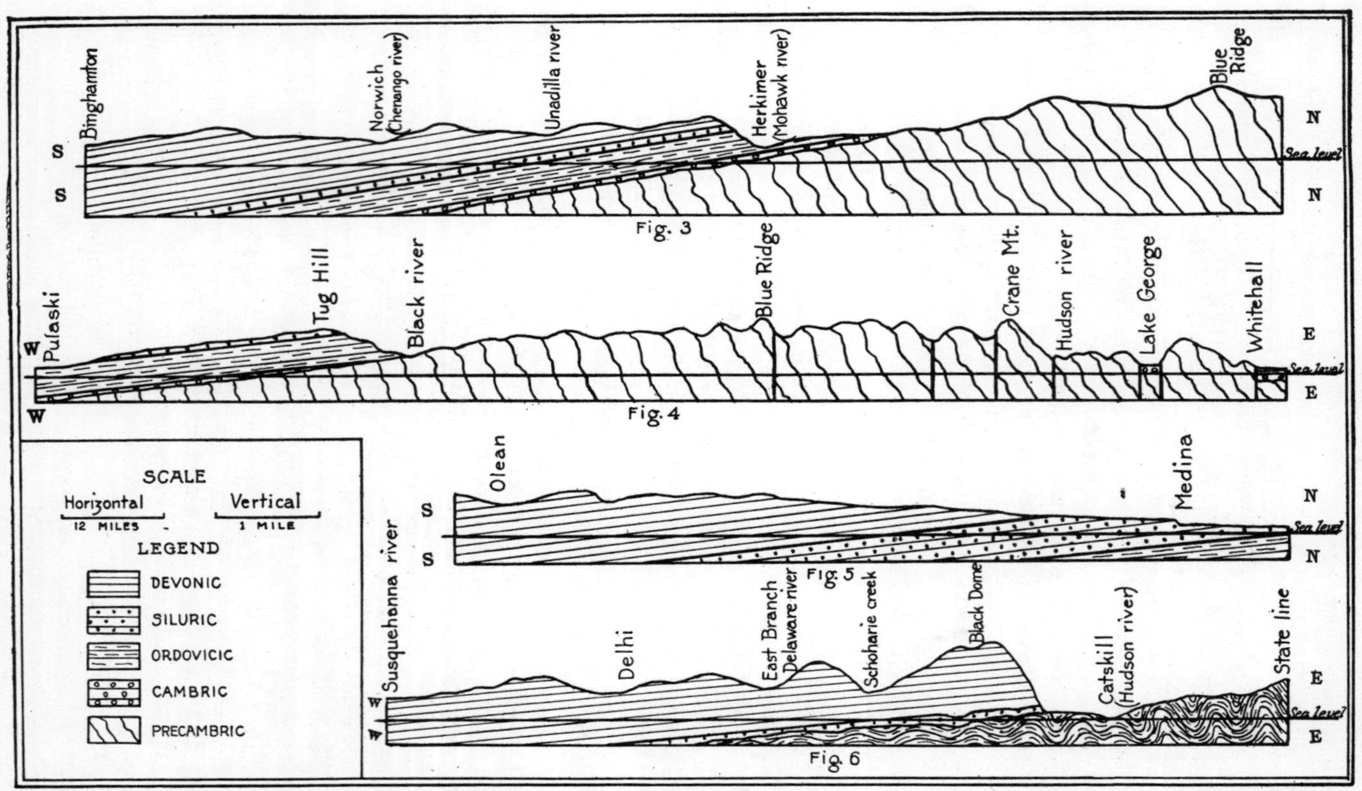


FIG. 3–6 Generalized structure sections through various parts of New York State, showing the attitude and relations of the great rock masses.

plateau and Mohawk valley provinces where the hard limestone lies at an altitude of more than 1000 feet, and directly overlies the soft shales of the valley whose altitude is only a few hundred feet.

## CATSKILL MOUNTAIN PROVINCE

This is the most rugged of all provinces in the State and, next to the Adirondacks, contains the greatest elevations. Slide mountain (4205 feet) is the highest, while a number of points range from 3500 to over 4000 feet in altitude.

The rocks are all of Devonic age and consist almost entirely of sandstones and conglomerates. Except for a slight westward undulation, these rocks are arranged in practically horizontal layers and show an aggregate thickness of several thousand feet (see figure 6). Lying under these Devonic rocks and outcropping at the very base of the mountains on the north and east, are various formations of Siluric age.

The term " mountains " as applied to the Catskills requires some explanation. The more typical mountains of the world have been formed by folding or faulting of the strata, or by igneous activity, or by two or all of these causes combined. For example, in the development of the Appalachians both folding and faulting have played prominent parts, while in mountains like the Sierras or Adirondacks, folding, faulting, and igneous action have all been important. The Catskills, however, in which these typical mountain phenomena are wholly lacking, are to be properly placed in the category of what we may call " erosion mountains." Mountains of the pure erosion type are due to an uplift of land high above sea level, followed by deep dissection of the elevated mass by the action of streams. The Catskills are only an easterly extension of the plateau province where the rocks are more resistant and perhaps the elevation of the region was greater, so that the streams were able to cut deeper trenches while the harder rocks of the divides have so far prevented a general wearing down of the region. The Catskills furnish a remarkable example of a high plateau deeply dissected by numerous streams. The whole topography is very rugged, all being much like that of the highest portion of the Adirondacks around Mount Marcy (compare plates 2 and 5). The Catskills, however, lacking the proper structural features, show practically no tendency to parallel arrangement of ridges or mountains as is so common in the Adirondacks.

On the south the Catskill province almost grades into the folded region of the Appalachians, while on the west it gradually merges into the southwestern plateau. On the north the Helderberg

escarpment, standing out abruptly and to a great height, forms a sharp boundary. On the eastern side the Catskills present a very steep, high front facing the Hudson valley. This steep front rises about 3000 feet and consists of hard Devonic sandstones and conglomerates overlying the Siluric strata (see figure 6)

## MOHAWK VALLEY PROVINCE

The Mohawk valley province, though comparatively small, is of great importance because it so clearly separates the Adirondack highlands on the north from the highlands of the Catskills and southwestern plateau provinces on the south. In fact it should be noted that the Mohawk valley is by far the lowest passageway across the mountains between the St Lawrence river and the southern end of the Appalachian range. This low pass is one of the great eastern "gateways" which, with the St Lawrence, have afforded the easiest means of communication between the Atlantic seaboard and the region west of the Appalachian mountains.

The comparatively narrow inner valley through which the river now flows is often erroneously called the Mohawk valley, but in reality the whole depression, from 10 to 30 miles wide and fully 1000 feet deep, between the northern and southern highlands of the State, should be called the Mohawk valley. At Little Falls the inner valley narrows to a gorge several hundred feet deep, where the river has cut its way through a preglacial divide (see plate 7 and figure 7). Had it not been for the recent cutting of this gorge (see explanation accompanying plate 44 in chapter 6) through the barrier at Little Falls, the Mohawk valley would never have been so important as a great gateway between the Atlantic coast and the west. Today the four tracks of the New York Central Railroad, two tracks of the West Shore Railroad, the Erie canal (now being enlarged to the Barge canal), an important highway, many telegraph and telephone wires, and the Mohawk river all pass through this narrow gorge and within a few hundred feet of sea level. Eastward and westward from Little Falls, the inner valley is generally fairly wide and open (see plate 7). At Little Falls the Mohawk river is less than 400 feet above sea level and even at Rome, in the western part of the province, the river shows an altitude of only 420 feet.

The principal rocks of the province are shales, sandstones and limestones of Cambric and Ordovicic ages; of these the soft, black shales of Trenton, Utica, and Frankfort ages are in greatest abundance. The valley owes its existence largely to the presence of this belt of soft shales lying between the hard crystalline rocks

of the Adirondacks on the north and the comparatively hard limestones immediately southward. The work of erosion has made rapid progress in this belt of weak rocks, and at two places, Little

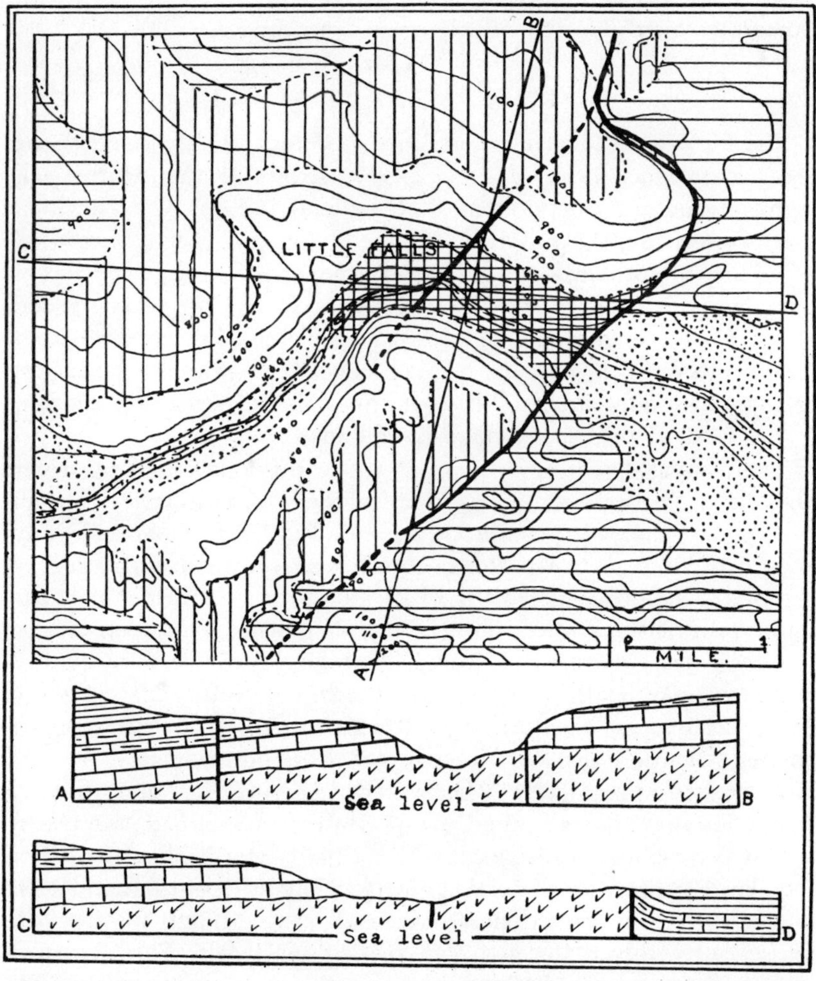


FIG. 7 Geologic and topographic sketch map and structure sections of the vicinity of Little Falls. Crosslined area at bottom of gorge = Precambric rock (syenite); blank areas = Little Falls (Cambric) dolomite; vertical line areas = Trenton (Ordovicic) limestone and shale; horizontal lined areas = Utica (Ordovicic) shale; dotted areas = Quaternary sand and gravel. Heavy black lines are faults. The structure sections show the condition of things along the lines AB and CD. In the sections the vertical scale is four times exaggerated.

Based upon map by H. P. CUSHING

Falls and "The Noses" (Yosts), the river has cut down to the Precambric (Adirondack) rock. In general, the rock formations of the province tilt slightly southwestward and show folding only on a very small scale. From Little Falls eastward, however, the strata are greatly disturbed by numerous nearly north-south faults which are often of considerable magnitude (see figure 25).

## ERIE-ONTARIO PLAINS PROVINCE

On the extreme western side of the State, and lying between Lake Erie and the Southwestern plateau, there is a strip of land only a few miles wide which may be called the Erie plain. This plain is of very low relief and slopes from an altitude of from 800 to 900 feet down to the level of Lake Erie, whose altitude is 573 feet. Where the Erie plain joins the Southwestern plateau there is a very decided change of slope. The rocks underlying this plain are dark shales of Devonic age and show the usual slight southwestward tilt (see plate 31).

The Ontario plain is much larger and lies between Lake Ontario and the Southwestern plateau, the southern boundary being marked by the "Helderberg escarpment." This large province slopes gradually to the shores of Lake Ontario and is remarkably free from relief features of any considerable magnitude. One that deserves mention is the presence of many hundreds of low, glacial knobs (drumlins) which are thickly scattered over the whole plain between Rochester and Syracuse (see plate 42). Another feature which serves to break the monotony of the plain on the west is the low but distinct escarpment of Niagara limestone, which extends from Lewiston eastward to beyond Lockport. On the east the Ontario plain gradually merges into the Mohawk valley province on the one hand, and on the other hand comes against the western foot of the highlands of the Tug Hill province.

The rocks underlying the Ontario plain are chiefly sandstones, limestones and shales of Siluric age, which show the usual tilt toward the south. At the extreme northeast, limestone and shale of Ordovicic age are present and these show a slight westward tilt.

## TUG HILL PROVINCE

The Tug Hill region is worthy of recognition as a distinct physiographic province because we have here a highland mass of considerable extent entirely separated from the neighboring provinces. The highest point, six miles west-northwest of Lyons Falls, Lewis

county, reaches an altitude of nearly 2100 feet, while the central portion of the province, covering many square miles, is remarkably flat and swampy with the general level above 1800 feet of elevation. On a smaller scale, this is as truly a plateau as the great Southwestern plateau already described and, interesting to note, this Tug Hill plateau is merely an erosion remnant of the great upraised Cretacic peneplain (see chapter 5) which formerly included all of New York State.

On the south and west this province slopes rapidly downward to the lowlands of the Mohawk valley and Ontario plain provinces, while on the east and north the Black river valley sharply separates this province from the Adirondack and St Lawrence Valley provinces. The rapid descent into the Black river valley bottom is everywhere 1000 feet or more over a series of high, steep terrace fronts (see plate 8). In passing, it should be stated that, though seldom recognized, this Black river depression takes rank as one of the few greatest valleys within the borders of the State.

Near Boonville, and at an elevation of about 1100 feet, occurs the division of drainage between the Mohawk and Black rivers, and this divide forms the highest land connecting the Tug Hill and Adirondack provinces. But in spite of this partial connection and the close proximity of the province to the Adirondacks, the rock formations and structure are wholly different from those of the Adirondacks while they greatly resemble those of the Southwestern plateau. The rocks are all of lower Paleozoic (chiefly Ordovicic) age, with several hundred feet of limestone at the base followed by about a thousand feet of shales, the whole being capped by a resistant sandstone of Siluric age. These strata tilt slightly westward but they have never been disturbed by folding, faulting or igneous activity (see figure 35).

## ST LAWRENCE VALLEY PROVINCE

The St Lawrence valley, lying along the northern boundary of the State, is a great, open depression of comparatively simple structure and near sea level. Where the river leaves Lake Ontario, the elevation is only 247 feet, while points with elevations more than a few hundred feet seldom occur within the province. As shown on the accompanying map (plate 9), low hills are common over the valley floor. The Thousand Islands form a remarkable physiographic feature of the province, where the wide, slow-moving St Lawrence river does not occupy any very distinct channel, but

rather flows across a broad, low, hilly region of very moderate relief thus allowing the low rocky hills to stand out as islands.

The rocks are chiefly sandstones and limestones of Cambric and Ordovicic ages, though, in the vicinity of the Thousand Islands numerous patches of the underlying Precambric (Adirondack) rocks are exposed as on many of the islands themselves. Folds, faults and igneous rock are not present except in the Precambric rocks, and the strata may be regarded as a comparatively thin mantle of nearly horizontal layers overlying the Precambric rocks.

## CHAMPLAIN VALLEY PROVINCE

The Champlain valley bounds the Adirondacks on the east and the province should be regarded as a great depression separating the Adirondacks on the west from the Green mountains on the east. Much of the valley bottom is filled by the waters of Lake Champlain (elevation 101 feet). Along the western shores of the lake the topography is characteristically hilly, though seldom above 500 feet in elevation. The transition to the higher and rugged Adirondacks is generally rapid.

The rocks occupying the valley bottom are sandstones, limestones and shales of Cambric and Ordovicic ages. These formations are much disturbed by numerous faults, often of considerable magnitude, and in fact there is good reason to think that the whole Champlain valley is of the nature of a great fault-trough or depression.

## HUDSON VALLEY PROVINCE

**General description.** Looked upon in a broad way, the Hudson valley province is a depression lying between the western highlands of New England and the eastern highlands of New York. Well toward the south the true valley feature is somewhat interfered with by the presence of such elevated masses as Shawangunk mountain and the Highlands-of-the-Hudson. A very detailed classification of topographic features would call for four or five provinces instead of the one here called the Hudson valley province. Since even this southern part, however, is lowland compared with the Catskill mountains immediately westward and since the rock structures are so similar and characteristic throughout the region, though the kinds of rocks vary considerably, it seems best for our purpose to treat all together as the Hudson valley province and then very briefly describe each of the minor subdivisions of the province.

The foundation rocks of the whole province comprise various formations of Precambric, Cambric, and Ordovicic ages, while in a few places mere surface layers of Siluric, Devonic, and Mesozoic strata occur (see geologic map, figure 1). All the rocks, except these few younger surface layers, are highly folded, which constitutes the most characteristic structural feature of the whole province. In fact, as is hereafter shown (see chapter 4), there are here exposed the roots or remnants of a portion of the great and very ancient Taconic mountain range which at one time occupied this

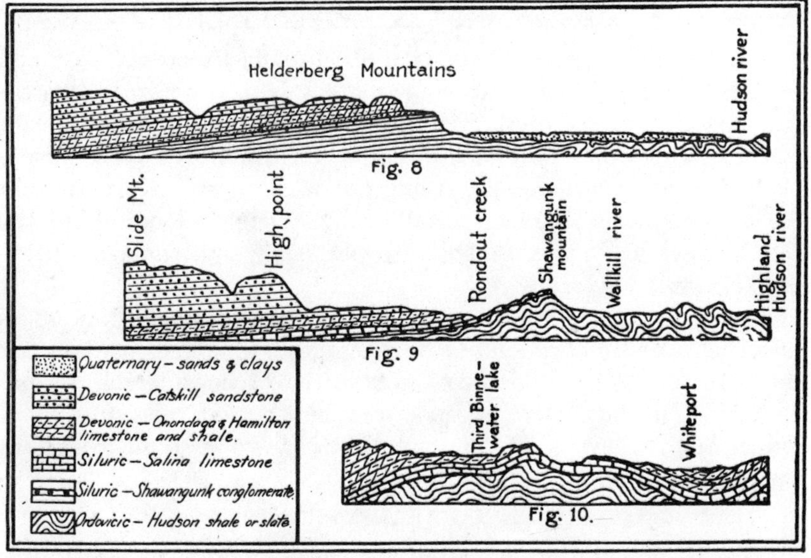


FIG. 8 Section from southwest to northeast through Albany county and showing the Taconic folds near the Hudson.

FIG. 9 Section from northwest to southeast across Ulster county and passing through Slide mountain and Highland village. Taconic and Appalachian folds both sides, as well as the structure of Shawangunk mountain.

FIG. 10 More detailed section through cement district at Whiteport, Ulster county. Both Taconic and Appalachian folds are well exhibited.

These sections all modified after Darton, N. Y. State Mus. Rep't 47, 1894, pp. 430, 490, 532

region. Thus from the geologic standpoint, the term Taconic province would be appropriate. All along the border of the province, as well as throughout the Hudson Highlands, the rocks are rather severely metamorphosed.

**Highlands-of-the-Hudson.** The Hudson Highlands extend across the Hudson valley in a northeast-southwest direction, and cover southern Orange county and northern Rockland county, and the region from southern Dutchess southward across Putnam

and into Westchester county. The relief is rather rugged with the higher points commonly reaching altitudes of over 1000 feet. The rocks are chiefly granites and gneisses of Precambric age, and are in most ways much like those of the Adirondacks.

**Region north of the Highlands.** North of the Highlands the rocks are in the main highly metamorphosed shales, sandstones and limestones, and the relief is generally low except along the eastern border where it is almost mountainous. A characteristic feature along this eastern side is the presence of long, fairly high, nearly north-south ridges separated by comparatively narrow valleys.

**Shawangunk mountain.** Lying close to the southeastern border of the Catskills and extending northeastward from the State line in Orange county well into Ulster county, is a distinct mountain ridge known as Shawangunk mountain. This long, narrow mountain rises 1000 feet or more above the surrounding country, and with the deep narrow Rondout valley immediately on its west side and the broad, open Wallkill valley on the east, it is truly a remarkable topographic form. The capping of very hard Siluric conglomerate upon the soft Ordovicic shales has caused the ridge to stand out so boldly against erosion (see figure 9 and plate 28).

**Region south of the Highlands.** Southern Rockland county is covered by Mesozoic (Triassic) sandstone. This rock is not folded but contains within its mass great sheets of lava which outcrop to form the Palisades along and west of the Hudson (see figure 20).

In southern Westchester county and in New York county there are highly folded and metamorphosed Precambric and Ordovicic rocks, and the country is typically hilly.

## LONG ISLAND PROVINCE

This province, including Staten Island, is really a part of the broad Atlantic coastal plain and is therefore practically devoid of any hard rock formations at the surface. Except for a few exposures of Cretacic sands and gravels along the northern border, the whole province is made up of glacial sands and gravels. From the standpoint of surface relief the province is clearly divisible into two parts, a northern and a southern, which are sharply separated from each other (see plate 12). The northern part is characteristically hilly, the hills being of glacial (morainic) origin. The maximum elevation is less than 400 feet, while in general the hills are from 100 to 200 feet high. This line of hills ends abruptly about midway of the island (north-south) and the southern part of the province is a sand plain of remarkable smoothness with a gentle slope toward the ocean.

## DRAINAGE

Considered as a great watershed, New York State takes rank as one of the most noteworthy in the United States. The waters of the State, except for a little in the southeast, enter the sea at five widely separated places, namely, Gulf of St Lawrence, New York bay, Delaware bay, Chesapeake bay, and the Gulf of Mexico, through the five well-known rivers, namely, St Lawrence, Hudson, Delaware, Susquehanna, and Mississippi (through the Allegheny and Ohio rivers).

**Mohawk-Hudson basin.** The principal stream of this, the largest drainage basin of the State, is the Hudson river which is especially noteworthy in two ways, first because it is by far the largest stream whose course is wholly within the borders of the State, and second because soon after emerging from the Adirondacks (near Glens Falls), its course, for nearly 200 miles to its mouth, is remarkably straight in spite of the fact that it traverses the principal structural lines of a highly folded and disturbed region. Its apparently anomalous, deep, narrow, granite-walled channel across the Highlands of the Hudson (see plate 10) will be explained in chapter 6. The chief tributary and essential part of the Hudson, the Mohawk river, has its headwaters in the very center of the State and reaches the Hudson after flowing eastward for more than 100 miles.

The drainage of the southern Adirondacks, even as far north as Mount Marcy, and of the eastern Catskills passes into the Hudson river. Except for very minor contributions from the edge of New England and New Jersey, the whole river system derives its water from within the boundaries of the State.

**St Lawrence basin.** This drainage basin comprises all the northwestern Adirondacks and reaches well into the heart of the mountains. All the larger streams, which are of very moderate size, flow northwestward in remarkably parallel courses until, emerging upon the floor of the St Lawrence valley, they swing around to northeasterly courses and generally flow for a good many miles parallel to the great river itself before entering it. This latter phenomenon is, no doubt, to be explained on the basis of topographic changes due to the great Ice age. The largest and longest stream in the basin is the Raquette river which, after a devious course of more than 100 miles, including passage through two or three large lakes, enters the St Lawrence at the northern State boundary.

**Ontario basin.** All the streams of this basin enter Lake Ontario and pursue courses that, if continued, would tend to converge at a

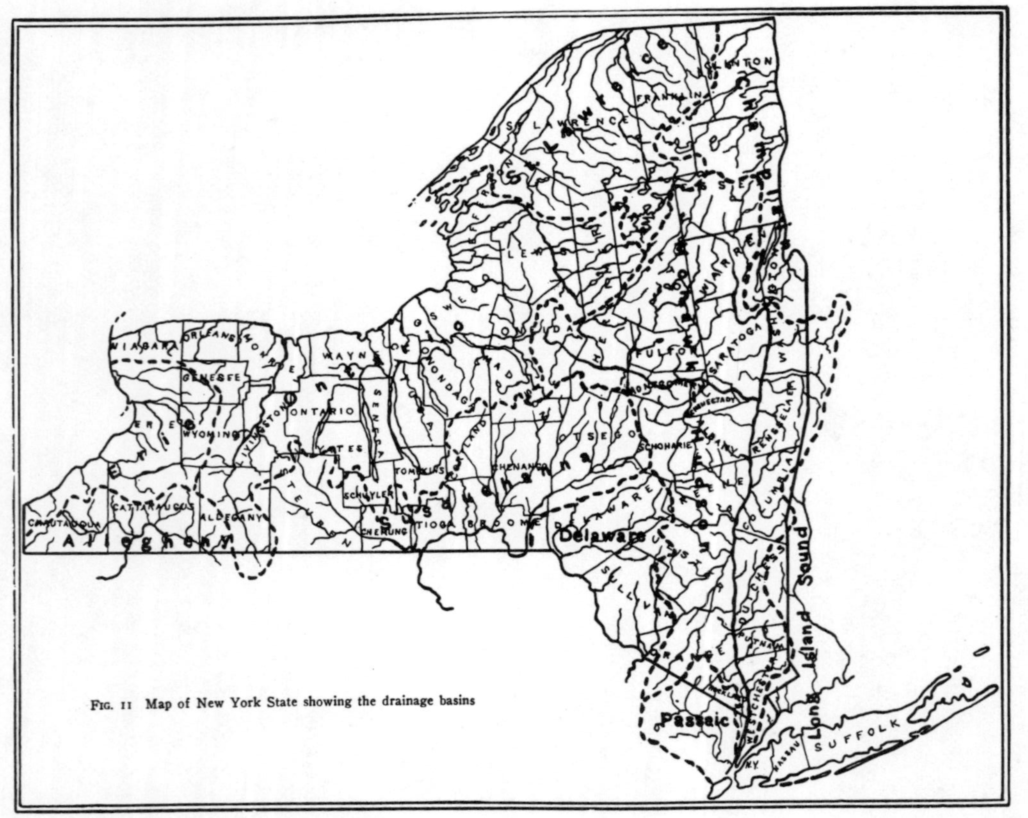

FIG. 11 Map of New York State showing the drainage basins

point in the lake basin. Thus the streams in the western portion of this drainage basin flow north to northeastward, northward in the middle portion, and westward to northwestward in the eastern portion. Three rivers should be mentioned: the Genesee, with its source in the highlands of Pennsylvania, flows northward across the entire southwestern plateau; the Oswego, toward the middle of the basin, takes one arm (Seneca river) from the west to drain the large Finger lakes and another arm from the east to drain Oneida lake and part of the Tug Hill plateau; Black river drains the extreme eastern portion of the basin, and a number of prominent tributaries flow southwestward from well within the Adirondacks to join the main stream in the great Black river valley.

**Susquehanna basin.** All the waters of the Susquehanna river are derived from the Southwestern plateau. The main stream (within New York) flows southwestward and together with its numerous large tributaries drains much of the eastern portion of the plateau region, especially in Otsego county. A number of the tributaries rise on the very crest of the Helderberg escarpment and within a few miles of the Mohawk river. The Mohawk river is fully a thousand feet below the crest of the escarpment, and its course is at right angles to that of Susquehanna tributaries.

The Chemung is the principal tributary from the northwest and drains a good portion of the south-central plateau.

**Other drainage basins.** The Champlain basin comprises the eastern border of the Adirondacks, including some of the highest and most rugged of those mountains. Nearly all the streams are short and most of them descend eastward very rapidly into Lake Champlain.

The Delaware, by means of its upper waters, drains the western and southern Catskills and the main stream, after flowing southeastward along the State line between New York and Pennsylvania, from Delaware county to Orange county, suddenly swings southwestward to pass through the famous Delaware water gap in the Kitatinny range.

The Allegheny river sends out a number of small branching arms to drain the extreme southwestern portion of the Southwestern plateau. Chautauqua lake (elevation 1338 feet) which lies at the very western edge of the plateau and close to Lake Erie, has its outlet into the Allegheny.

The Erie basin contains no river of much consequence, the small streams all flowing westward or northwestward across the narrow Erie plain and into Lake Erie or the Niagara river. Nearly all

these streams rise along the western edge of the Southwestern plateau.

In quantity very little of the water of the State enters the Long Island sound basin, a few small streams north of the sound together with the numerous but very small streams of Long Island comprising the whole supply.

The Passaic basin is mostly confined to New Jersey, with a few small streams having sources in Rockland and Orange counties.

## Chapter 3

# PRECAMBRIC HISTORY

### THE GRENVILLE FORMATION

In the Adirondack mountains, and also probably in the Highlands-of-the-Hudson, we have the earliest known records of the physical history of New York State. These records are written in a series of rocks named the Grenville, so called from a town in Canada where the rocks were first well known. The Grenville is of interest, not only because it is the most ancient rock formation so far discovered in New York, but also because it takes rank among the very oldest rock formations of the earth.

Until about fifteen or twenty years ago the real significance of the Grenville and its closely associated rocks in the Adirondacks was not recognized, but now many of the leading events of that very early history are established. As in human history, so in earth history, the earliest records are the most obscure and difficult to read and, in the one as in the other, it is easy to pass from conclusions properly based upon facts to mere speculations. Many problems regarding the Precambric history of our State yet remain to be solved, but in these pages it is rather the purpose to describe only those historical events which have been well established.

The Grenville consists of a great series of marine water-laid rocks which are clearly older than the Paleozoic because these latter rest upon the Grenville in many places. As will be shown below, the Grenville strata have been so profoundly changed from their original condition that certain of the highly sedimentary features have been obliterated. Thus the absence of water-worn particles and fossil shells, both of which are so characteristic of sedimentary deposits, is due to complete crystallization (metamorphism) of the Grenville strata since their formation. Nevertheless we have certain proofs of the sedimentary origin of the Grenville. The fact that these rocks commonly occur in alternating layers, which stand out in sharp contrast because of marked difference in composition and color, furnishes strong evidence that this distinct banded effect is due to differences in original sedimentation. A great mass of igneous rock is generally characterized by homogeneity throughout; a mass of typical sediments, on the other hand, is arranged in distinct layers, such as shale, sandstone, or limestone which show frequent differences in composition. In the Grenville, especially of

the northwestern and southeastern Adirondacks, there are extensive beds of crystalline limestone, such as the Gouverneur marble of St Lawrence county. Such rocks could not have been of igneous origin. In many places, and sometimes in sharp contact with the limestone, are beds of almost pure quartz rock, which are certainly not igneous but which represent original sandstone layers. Again, there are very extensive deposits throughout the Adirondacks of generally darker colored rocks rich in such minerals as quartz, feldspar, garnet, mica, pyroxene, and amphibole. These rocks, because of their constant close association with the strata just described as well as their banded structure, are also clearly ancient sediments. The composition of these latter rocks shows that the original sediments were muds, often with sand or lime. Another argument in support of the sedimentary character of the Grenville is the presence of flakes of graphite (plumbago) which are so commonly disseminated throughout the formation. In some places the strata are so filled with graphite that the mineral is mined, as in Essex and Saratoga counties. Carbon existing under such conditions is probably of organic origin and represents the final stage in the decomposition of organisms which lived in the waters while the Grenville strata were being deposited. The occurrence of so much garnet is also at least highly suggestive because this mineral is especially common in crystallized sediments in many parts of the world.

*Having established the sedimentary origin of the oldest known formation in New York State, we are led to the interesting and important conclusion that this Grenville formation is not the oldest which ever existed in the State.* The Grenville sediments must have been deposited, layer upon layer, upon a surface of still older rocks. A knowledge of the character and composition of such Pregrenville rocks would be of very great interest, but thus far we have no positive evidence that such rocks are visible in the Adirondacks, although certain rocks still of somewhat doubtful age and origin may belong to that very ancient rock floor. Again, the fact that Grenville sediments were being deposited under water carries with it the corollary that there must have been land somewhere at no great distance from the area of deposition because then, as now, such sediments as muds and sands could have been derived only from the erosion or wearing away of land and have been deposited in great sheets one above the other, under water adjacent to the land mass. Here, too, we are as yet utterly in the dark so far as any knowledge of the location or character of that very ancient land is concerned.

## DISTRIBUTION, THICKNESS AND AGE OF THE GRENVILLE

The Grenville is associated with rocks of younger age in the Adirondack region, so that the formation is not present as a single, continuous mass of surface rock covering the whole area. It is, however, so abundant and widespread in great and small areas throughout the Adirondack province that we may confidently assert that this whole district was under water during Grenville time. As the geologic structure strongly suggests and as certain deep wells prove, rocks of this age must extend, under cover of the later Paleozoic sediments, for a considerable distance beyond the Adirondack area. Precambric rocks have long been recognized in the Highlands district of the lower Hudson and recent work makes it practically certain that strata of Grenville age exist there. Precambric (doubtless including Grenville) occurs along the western border of New England and it should also be mentioned that Grenville strata are extensive over much of southeastern Canada. Presence of Grenville strata in southwestern New York is somewhat doubtful because, if there, they are effectually concealed under the heavy cover of Paleozoic strata. The positive existence of Grenville, however, just to the north in Canada and in northern and southeastern New York, makes it more than likely that the Grenville underlies the Paleozoic rocks of western New York also. Such a widespread distribution of the Grenville sediments shows that deposition went on in a very large body of water; large enough, in fact, to be called an ocean. *Thus, bearing in mind all the facts, we are led to the important conclusion that, during Grenville time, all of northern and eastern and probably southwestern New York was under the sea. In other words, the most ancient known geographic condition in New York was a great expanse of ocean water covering most, if not all, of the State.*

As the rocks are badly disturbed and folded, and as the top or bottom of the formation has never been recognized as such, it is impossible to give anything like an exact figure for the thickness of the Grenville rocks. Continuous successions of strata have been observed in enough places, however, to make it certain that Grenville strata were piled, one layer upon another, to a thickness of many thousands of feet. This clearly implies that the Grenville ocean existed for a vast length of time which must be measured by no less than a few million years, because in the light of all our knowledge regarding the rate of deposition of sediments, such a very long time was necessary for the accumulation of so thick a mass of rocks. It does not necessarily follow that the Grenville

ocean was thousands of feet deep when the deposition began, because there may well have been a gradual subsidence of the sea floor during the process of sedimentation, which means that there was not necessarily very deep water at any time. In fact, the very character of the sediments clearly indicates that the Grenville ocean was, for most part at least, of shallow water, for such sediments as sands and muds have rarely if ever been carried far out into an ocean of deep water. The great ocean abysses of today are not receiving any appreciable amount of land-derived sediments. Hence it is practically certain that the very ancient Grenville sea bottom gradually settled as the sediments accumulated. Similar phenomena are definitely known to have occurred in many later basins of deposition.

The reader may naturally be disposed to ask, How long ago did the Grenville ocean exist? There are grave difficulties in the way of answering this question in terms of years, as we have nothing like an exact standard of this kind for comparison. While it is fully recognized that not even approximate figures can be given, a very conservative statement would ascribe an age of twenty to twenty-five million years to the Grenville strata. Whatever its exact duration may have been, the time is utterly inconceivable to us, and the important thing to bear in mind is that the great events of earth history which have transpired since that time require a lapse of many million years as shown by the enormous accumulations of sediments in many parts of the earth, and by the building up and wearing away of one great mountain range after another. The reader will better appreciate the significance of these statements after he has studied the following pages. In the table of geological time divisions given in chapter 1, the two oldest periods are the Archean and Algonkian respectively. The Grenville can not with certainty, as yet, be placed in either of these periods, although according to the best evidence it should be classed with the Archean. In the meantime it is advisable to refer to all formations older than the Paleozoic simply as Precambric.

## LIFE IN THE GRENVILLE OCEAN

All that can be said regarding the life of the Grenville ocean is that it existed as indicated by the presence of the graphite. Although we can not even state whether the organisms were plant or animal, the fact that there was life in that very ancient ocean is a matter of no little significance. Anthracite coal, which is chemically

very similar to graphite, occurs in the Carbonic strata of Pennsylvania, and is derived from plants through the process of carbonization. Graphitic anthracite of like origin occurs in a smaller way in Rhode Island. Hence it seems likely that the graphite of the Grenville represents the remains of plants, probably of the seaweed type since there is much evidence against the view that any of the higher land plants existed at that very early time. This by no means proves the absence of animals from the Grenville ocean, because animals with only soft parts would have left no record, while calcareous or silicious shells would doubtless have been recrystallized by the severe processes of metamorphism to which these rocks have been subjected.

## EARLY PRECAMBRIC IGNEOUS ACTIVITY

*After the accumulation of the Grenville sediments, igneous activity, took place on a large scale, when great masses of molten rock were pushed or intruded into the sediments from below.* Several different times of igneous activity have been definitely recognized and the general effect of the great invasions of molten rocks was to break the Grenville up into patches. In many cases considerable masses of Grenville were pushed aside or displaced by the molten masses while, to a greater or less extent, there may have been an actual melting in or assimilation of Grenville rocks by the molten intrusions. As we have already learned, igneous rocks are those which have cooled from a molten condition, and of these there are two important types, called respectively, plutonic and volcanic. Both of these types of igneous rocks are found in the Adirondacks, but the plutonics are by far the more prominently developed.

So far as we know, the first great intrusion of molten rock in the Adirondacks is represented by the present large area of so-called anorthosite in Essex and Franklin counties. This is a very coarse-grained, plutonic rock of bluish gray color when fresh and consists chiefly of a feldspar (labradorite). The intrusion was practically confined to a single area comprising about 1200 square miles. That this rock is younger than the Grenville is demonstrated by the fact that tongues of the anorthosite have been observed cutting through the Grenville (see figure 12). This intrusion differs from the later intrusions in that it was practically a single great mass which broke its way through the Grenville in but one place in the whole Adirondack region. In a few cases small patches of Grenville were caught in the molten flood and may now be seen within the anorthosite mass. For the most part, however, the molten rocks

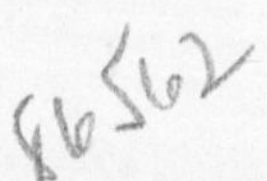

either completely pushed aside the Grenville or melted it into itself, the former being the more likely. Many of the highest points within the Adirondacks, like Mt Marcy, are in this anorthosite area.

The next clearly recorded event, after the anorthosite intrusion, was very widespread igneous activity when the rocks of the granite-syenite series, now so well known in the Adirondacks, were forced upward into the Grenville sediments. To be precise, there were at least two or three periods of intrusions of such rocks, the oldest probably being represented by the so-called Laurentian granite of the Thousand Islands region. For our purpose it will suffice to regard these as having all been intruded at about the same time, since the sum total of effects is much the same as though there had been but a single period of activity. Granite is a plutonic, igneous rock which consists essentially of quartz and feldspar (orthoclase),

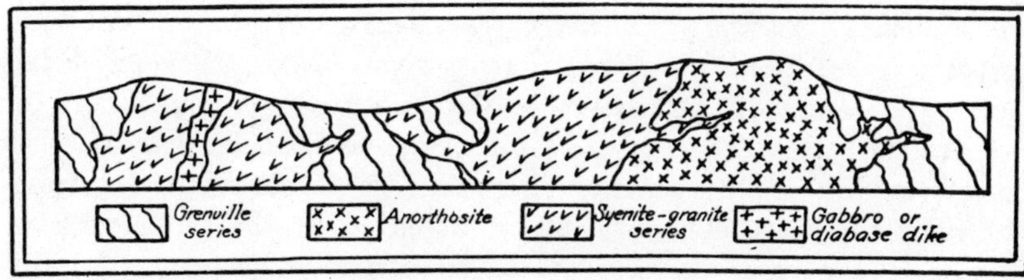


FIG. 12 Generalized section showing the relations of the most common Precambric (Adirondack) rocks to each other and how their relative ages are determined.

together with more or less black mica, hornblende, or augite. Syenite is the same except that quartz is much less prominent or lacking. The fresh rock is of a greenish gray or pinkish gray color, while on weathered surfaces the color is usually light brown. Many of the highest mountains outside the anorthosite area are of granite or syenite.

The present distribution of these rocks shows that the molten masses broke into the Grenville in very irregular fashion, sometimes pushing the Grenville aside; sometimes enveloping great or small masses within the molten flood; or, in other cases, apparently leaving large Grenville masses intact. All portions of the Adirondacks felt the force of the intrusions and a detailed geologic map of the region would show a decided patchwork effect (see figure 13) due to the irregular manner in which the Grenville has been cut up by the igneous rocks. These igneous rocks are generally easily distinguished from the old sediments because of their homogeneity in

large masses and lack of sharply defined bands of varying composition.

That these granite-syenite rocks are younger than the anorthosite is demonstrated by the fact that tongues of the former have been observed to cut the latter (see figure 12). Since the granite and syenite now visible in the Adirondacks are plutonic rocks, they

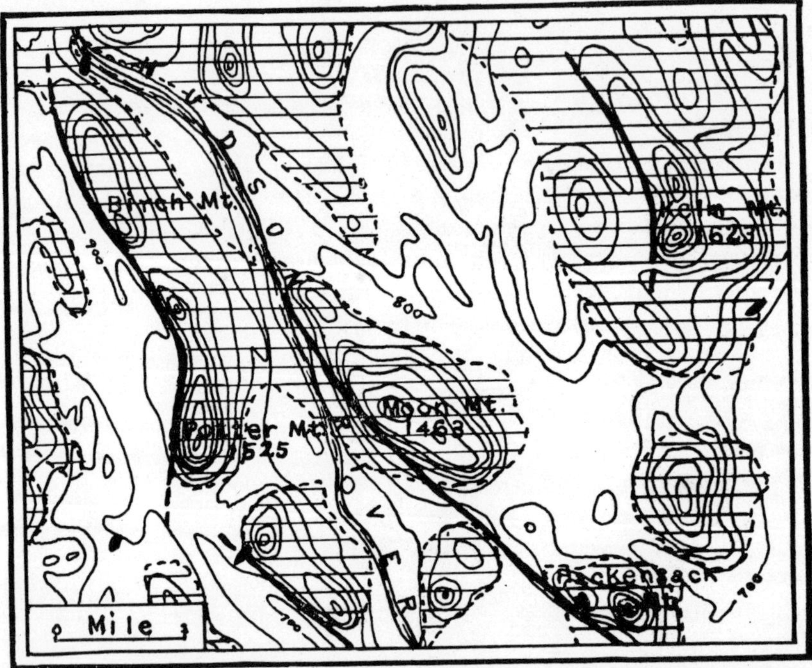


Fig. 13 Geologic and topographic sketch map of the southeastern portion of the North Creek sheet (Warren county), showing the surface relations of the common Precambric rocks in the southeastern Adirondacks. Contour interval 100 feet. Horizontally lined areas = syenite or granite; blank areas = chiefly Grenville; small heavy black lines are faults. The faults here shown are minor ones which do not follow the NE-SW trend of the major faults of the eastern Adirondacks. The so-called "patch-work" effect is well shown. Note how the more resistant rocks form the mountains which rise above the general level of the Grenville. (W. J. M.)

could not have reached the surface by intrusion, as such rocks can be formed only by slow cooling under great pressure thousands of feet below the earth's surface. They now appear at the surface because of vast removal, by erosion, of the overlying original materials.

It should be stated that this period of igneous activity is by no means confined to the Adirondacks. Similar intrusions are known in the Highlands-of-the-Hudson and in Canada. The covering of Paleozoic strata in southwestern New York prevents direct observation, but all things considered, it is more than likely that much or all of that part of the State, stripped of the Paleozoics, would also show Grenville cut up by granite and syenite.

## FOLDING OF THE ROCKS AND UPLIFT OF THE ADIRONDACKS

*At some time after these great periods of igneous activity and cooling of the rocks, the whole Adirondack region was subjected to an enormous pressure as a result of which the rocks were highly folded and compressed.* The Grenville strata now seldom lie horizontally but are tilted at all sorts of angles, and even the igneous rocks thus far described, show unmistakable evidence of having been greatly compressed because the minerals are flattened out or arranged in parallel fashion, often exhibiting a crude, banded structure. Rocks which have thus been changed are known as gneisses and may be either igneous or sedimentary.

Still later than the granites and syenites, there occurred intrusions of gabbros in the form of dikes or fissures filled with igneous rock. This gabbro is a very dark gray, rather coarse-grained, plutonic rock, and it has been found in numerous small masses throughout the Adirondacks, but especially in the eastern portion. Its age, younger than the granite-syenite series, is demonstrated by the fact that it often breaks through those rocks, and also because it is not nearly so much metamorphosed.

The Grenville sediments were completely crystallized as a result of this metamorphism so that all beds of limestone were converted into marble, sandstone into quartzite, and the shales into gneisses of varying character. Thus the Grenville strata have been very greatly altered from their original condition, which explains why they do not look like the more typical and familiar sediments of later age. The manner in which the Grenville strata, especially the limestones, were crumbled and folded shows that the rocks must have been in a more or less plastic condition when the pressure was exerted (see plate 13). Heat and moisture, no doubt, aided in this process which we call metamorphism. Such a process can take place only at thousands of feet below the surface of the earth where the rocks, under the enormous weight of overlying material, would act like

plastic masses when subjected to any great lateral pressure within the earth.[1]

Rocks at the surface of the earth, subjected to the same lateral pressure, would be broken or fractured instead of bent or folded. It is not the present purpose to discuss the origin of such pressure within the earth, but suffice it to say that it is somehow due to the shrinkage of the interior portion of the earth, and the tendency of the exterior (or crust) to accommodate itself to the shrinking interior, thus producing lateral thrust (pressure) in this exterior portion.

We are now ready to discuss the first known uplift of the whole Adirondack region above sea level, or what we may call the birth of the first known Adirondack mountains. As we have just learned, the very character and structure of the rocks now exposed in the region, show conclusively that they were at one time deeply buried under thousands of feet of overlying materials, and the inference is perfectly plain that those materials must have been removed by erosion. Extensive erosion of any land mass means that the land must be above sea level and thus we come to the important conclusion that *the great mass of Grenville sediments were upraised well above sea level.* Just when the great uplift occurred can not be positively stated, but if it was not during, or after the igneous intrusions, it must have been shortly before them. It is quite reasonable to believe that the same great force which caused a welling up of so much liquid rock might easily have caused a decided uplift of the whole region. Again, it is quite plausible that there may have been no great uplift until the development of the lateral pressure which caused the metamorphism of the rocks. Still another view is that a lateral pressure, once started, first caused a welling up, at different times, of igneous rock and then, after the cessation of the igneous activity, the same force continued to compress, fold and metamorphose the rocks. Whatever the actual history may have been, it is at least true that the sum total of all effects, as we now observe them, harmonizes well with the view last expressed. The intense folding and tilting of the strata show that the amount of uplift must have been very considerable as is the case in all typical mountain ranges. We can not, however, even state the approximate height of those very ancient Adirondack mountains. The fact

---

[1] Professor Adams of McGill University, Montreal, has recently proved experimentally that rocks, under great pressure, flow like plastic matter.

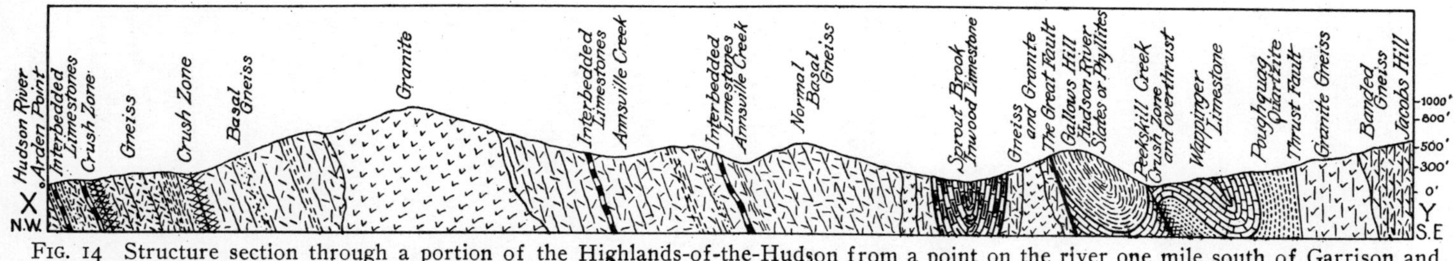


FIG. 14 Structure section through a portion of the Highlands-of-the-Hudson from a point on the river one mile south of Garrison and extending 6 miles southeasterly to Jacob's hill. The basal gneisses are highly metamorphosed and folded Precambric (Grenville?) sediments through which large masses of granite were erupted. Toward the right are closely infolded belts of early Paleozoic sediments.

After C. P. Berkey, N. Y. State Mus. Rep't 60, v. 2. 190[illegible], p. 371

that many thousands of feet of materials have been removed by erosion in order to expose the present rocks to view does not necessarily imply that the mountains at any time had so great a height because it is possible that, while elevation slowly progressed, material was steadily removed by the process of erosion. All our knowledge regarding later and better known mountains, however, leaves little doubt that those first Adirondacks were mountains vastly higher than those of today.

The same sort of uplift and succeeding profound erosion also affected the present region of the Hudson Highlands (see figure 14) and, this being the case, we can state with confidence that much, if not all, of northern and eastern New York was involved in this mountain-making process. For western New York the physical geography of this time has not been determined because all that region is so deeply buried under the Paleozoic strata.

## LATE PRECAMBRIC HISTORY

The fact that such a great thickness of rock was removed by erosion implies a vast length of time for the accomplishment of that work. According to all that we know regarding the rates of erosion of mountains of the present and past, the erosion of the Adirondacks must have extended over a period of at least several million years. When we consider that, at the opening of Upper Cambric time, most of the region had been worn down to the condition of a peneplain (see page 42), we are confident that no small amount of the erosion was accomplished even before the opening of the Paleozoic era, and that it continued well into the early Paleozoic. The whole problem of this later erosion and its effects will be treated in the next chapter.

*Well toward the end of Precambric time, igneous activity of a minor character took place in the formation of dikes* which, as we have learned, are fissures in the crust of the earth which have been filled with molten rock. In the Adirondack region there are several kinds of dike rocks, the most common being pegmatite and diabase. Pegmatite is a very coarse-grained, light colored rock of general granitic composition, the feldspar and quartz crystals often attaining lengths of several inches to a foot. Diabase is a fine-grained rock much like ordinary basaltic lava. These dikes are generally less than a mile long and comparatively narrow. That they are younger than the other igneous rocks of the region is abundantly proved by the fact that they cut through those rocks in many places

(see plate 17 and figure 12). That these dike rocks were intruded after the great pressure and uplift of the region is shown by the total absence of metamorphism or alteration of any kind along their contact lines. The fine-grained texture of these rocks, often with borders of glass, shows that they must have cooled close to the surface, and hence it is evident that most of the Precambric erosion of the region had been accomplished before the diabases were erupted. Such rocks suggest that there may have been volcanic activity at the surface but no positive proof for such activity can be given because, if such volcanic material ever existed, every trace of it has been removed by erosion. The diabases have been observed to cut the pegmatites and hence they are not only the younger of the two, but they must take rank as the youngest Precambric rocks in the State. In the Adirondacks the pegmatites are very widely distributed and common, while the diabases are most abundant in the northeast, less so in the northwest and southeast, and nearly absent from the southwest.

## Chapter 4

# PALEOZOIC HISTORY

## CAMBRIC PERIOD

In the preceding chapter we have seen that after the first known great Adirondack uplift the whole region, including the district of the Highlands-of-the-Hudson, was profoundly affected by erosion, and that this erosion began before the Paleozoic age and extended well into the early part of that era. Now the question may be fairly asked, What became of the sediments which were derived from the wearing down of the land during that vast length of time? We must admit that, in our present state of knowledge, we can not be certain as to what became of the Prepaleozoic sediments. They may have washed westward or southwestward into waters which might possibly have existed there; or they may have been carried northward or northwestward into Canada to help build up late Precambric deposits there; or they may have moved eastward toward or into the Atlantic basin. The question of the disposition of the early Paleozoic sediments, however, can be much more satisfactorily answered. Early and Middle Cambric deposits are extensively developed in the New England states and along the eastern border of New York State, including the Hudson Highlands region. Thus we have positive proof for the presence of the early and Middle Cambric sea over this region, and it is equally evident that much of this sediment which deposited in the sea was derived from the adjacent land masses in northern and eastern New York.

It was not until the opening of the Upper Cambric (Potsdam time) that any considerable portion of New York State was occupied by Paleozoic sea water. The fact that all the Cambric strata, including Lower, Middle, and Upper, are present in the New England country and along the eastern border of New York, while only the Upper Cambric is present in northern New York, clearly shows that the Cambric sea encroached upon the State from the east toward the west. To be more exact, it was probably from the northeast, because the greatest thickness of Upper Cambric strata is in Clinton county along the northeastern side of the Adirondacks.

The first deposit to form in the Cambric sea of northern New York was the Potsdam sandstone and the presence of this formation in the St Lawrence, Champlain, and lower Mohawk valleys

proves that the Potsdam sea occupied all these regions. Along the southwestern border of the Adirondacks the Potsdam is absent and there is not the slightest evidence that it ever was present, so that region must have been dry land in Potsdam time. In the southeastern Adirondacks the Potsdam sea certainly extended in as far as Wells (southern Hamilton county) and North River (northwestern Warren county), because small outlying masses of Potsdam sandstone occur at those places. These outlying masses were formerly connected with the larger areas but have become completely separated from them by extensive (Postpaleozoic) erosion and downfaulting. There is no evidence whatever that the sea covered the heart of the Adirondacks. *To summarize for northern New York, we may say that the Potsdam (Upper Cambric) sea covered the whole region except the central and southwestern Adirondacks which stood out as a great island in the midst of that ocean.* Southeastern New York certainly, and the middle eastern border of the State probably, were covered by the Upper Cambric sea, but whether that sea extended over the rest of the State has not been determined because all early Paleozoic strata, if present, are now deeply buried.

What do we know about the character of the topography of the land over which that ancient Potsdam sea spread? As a result of the very long erosion during late Precambric and early Paleozoic, thousands of feet of material had been removed so that rocks which had been so deeply buried were exposed at the surface, and the whole country must have been well worn down. Was the region worn down to the condition of a peneplain? Recent detailed studies on all sides of the Adirondacks furnish a very satisfactory answer to this question. In many places the Potsdam has been seen in actual contact with the Precambric rock whose surface oftentimes clearly proves that the whole region had reached a peneplain condition. Along the northeastern Adirondacks this peneplain was considerably rougher than along the northwestern and southwestern portions. This is explained by the fact that the northeastern area subsided first and consequently was not subject to wear quite so long as the latter named areas.

The accompanying figure (no. 15) affords an interesting example of the kind of peneplain topography here considered. It demonstrates that occasional low knobs of more resistant rock (for example, Grenville quartzite) protruded above the otherwise nearly featureless plain, because when the Potsdam sea overspread the

region such low knobs were not covered by the water. At this same locality there is a fine exhibition of coarse conglomerate at the base of the Potsdam sandstone, the boulders of the conglomerate often ranging from one to three feet across. These boulders were torn off the Adirondack cliffs by the waves of the Potsdam sea and were deposited near shore in local depressions of the old rock surface. The sandstone itself everywhere abounds in ripple marks, thus proving the shallow water (near shore) origin of the rock. All the rock in the walls of the famous Ausable Chasm (Clinton county) is Potsdam sandstone (plate 18).

Immediately overlying the Potsdam and showing about the same areal distribution, are alternating sandstones and limestone beds (Theresa formation) which show a thickness of from 50 to 200 feet. After still greater subsidence, the important formation known as the Little Falls dolomite (limestone) was deposited, layer upon

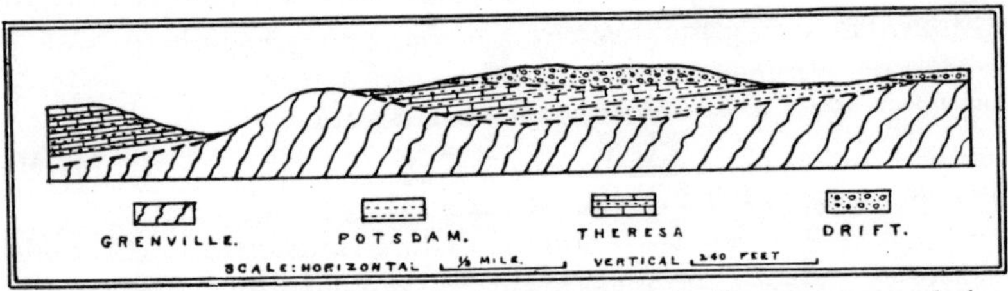


FIG. 15 Section passing through North Galway in Saratoga co. and showing how the Cambric (Potsdam and Theresa) strata overlap upon a hillock of Precambric rock. This knob of Precambric rock stood above the general level of the peneplain (early Paleozoic) and was not submerged under the Potsdam sea.

After W. J. Miller, N. Y. State Mus. Bul. 153, p. 5

layer, in the Upper Cambric sea. This formation which is hard, compact and of light gray color, shows a thickness of several hundred feet in the gorge at Little Falls. It rests directly upon the Precambric rock there (see figure 7), which shows that the Cambric sea spread over that region for the first time when that dolomite was forming. The Little Falls sea swept all around the Adirondacks, except what is now the western border from Trenton Falls to the Thousand islands district. Occurrence of the dolomite in small outlying masses at Wells (Hamilton county) and Schroon Lake (Essex county) proves that the Little Falls sea extended well into the eastern Adirondacks. Rocks of this age appear to be present in southeastern New York and if so, the Little Falls sea also overspread that region. Direct evidence for western and southern

New York is wholly lacking, but judging by the occurrence in central Pennsylvania of rocks of the same age as the Little Falls dolomite, it is highly probable that the Little Falls sea also covered western and southern New York.

The Little Falls dolomite is especially significant in two ways; first, because it is the youngest (uppermost) Cambric formation in the State, and second, there is a distinct unconformity at its summit. By unconformity here we mean that, after the deposition of the dolomite, at least all of northern New York was raised (without folding or faulting) above sea level and underwent erosion for a moderate length of time after which most of the region again settled below sea level to receive the deposits of later (Ordovicic) age. This old eroded surface and unconformity has been well established, and hence we learn that *the great Cambric period of the early Paleozoic era closed with all of northern New York, at least, well above sea level.* In southeastern New York, so far as known, the Cambric strata appear to grade into the Ordovicic and if so, that region was not raised above sea level at the close of the Cambric. We are wholly ignorant as to the physical geography of western and southern New York at the close of the Cambric because the records there are not accessible, as they are deeply buried.

## ORDOVICIC PERIOD

We are now ready to consider the physical condition of the State during the great Ordovicic period of earth history. During this time the Appalachian mountain region, the great Mississippi valley, and much of the far western region were almost continually under water (see figure 16). In fact, the very widespread distribution of thick Ordovicic strata shows that more of North America was covered by the Ordovicic sea than by any other sea, with the possible exception of the Precambric. Among the more prominent lands which persisted above water were Appalachia, a great land mass occupying what is now the Atlantic sea board and extending an unknown distance into the Atlantic, and another large land area in the Hudson Bay region of Canada. Sediments from those lands were washed into the Ordovicic sea which, for most part at least, covered the State during the entire period. In eastern and southeastern New York, the almost unbroken succession of Ordovicic strata shows that the sea was much of the time present there during the entire period. The prominent development of Ordovicic strata west and south of New York makes it practically certain

that the western and southern portions of the State were submerged under the Ordovicic sea. In northern New York, however, there is no positive evidence whatever that the whole Adirondack area was ever completely submerged during this period. Accordingly the central Adirondacks formed a persistent island in the Ordovicic sea. Furthermore, in northern New York there were various rather

FIG. 16 Generalized map of North America showing the relations of land and water during the Midordovicic period. Horizontally lined areas = land; blank areas = water. All of New York State was submerged except the central Adirondacks which stood out as an island. The conditions in Mexico and Central America are practically unknown.

local oscillations of level bringing the land around the island now above and now below sea level, but all such details are here omitted. For our purpose it will suffice to say that, except for the Adirondack island, northern New York was mostly below sea level during the Ordovicic. To summarize the above statement: *New York State*

*was completely submerged under the Ordovicic sea except for the Adirondack island and alternating land and water conditions immediately around that island.*

Without going into the details of the formations, it is important to note that the earlier Ordovicic deposits were almost wholly limestones, while the later deposits were nearly all shales and sandstones. Thus in southeastern New York the thick Wappinger limestone is overlain by the still thicker Hudson River shales and sandstones. In northern New York we have the Beekmantown, Chazy, Black River, and Trenton limestones overlain by the Trenton (Canajoharie), Utica, and Frankfort shales and sandstones. It should not be understood, however, that all the formations named are present in unbroken succession, because the oscillations of level (above mentioned) occasioned certain interruptions in sedimentation.

The predominance of limestone formation in the earlier Ordovicic sea of New York proves that the waters of that time were comparatively free from land-derived sediments and this, in turn, is best accounted for not by great depth of water and distance from land, but rather by the fact that all the nearest land areas were comparatively low and small, and hence were not undergoing very active erosion. During the later Ordovicic the adjacent lands were considerably higher and no doubt larger, so that vigorous erosion resulted and muds and sands were largely washed into the sea.

The aggregate thickness of Ordovicic strata in New York is between 2000 and 3000 feet. It should not be inferred from this fact that the Ordovicic sea was ever two or three thousand feet deep. Even the limestones abundantly show by ripple marks, mud cracks, fossils etc. that they were laid down in shallow sea water. The very character of the materials (old muds and sands) in the Upper Ordovicic formations shows that they could not have been deposited in deep ocean water. Such sediments are not now forming on the deep sea bottom. But how are these statements to be harmonized with the fact that nearly 3000 feet of Ordovicic strata exist in New York? During the whole period (with certain exceptions above noted) the land gradually subsided and in this slow downward movement stratum after stratum was formed upon the sinking sea-floor, so that at no time is it necessary to assume great depth of water. In general, the Ordovicic sea of North America must be thought of as a vast shallow (continental) sea which spread over most of the slowly subsiding continent. There were no ocean abysses at all comparable to those of the present ocean.

In the strata of Cambric age in New York, animal or plant remains are comparatively rare, while the Ordovicic rocks throughout fairly teem with fossils. If any single formation deserves special mention, it is the Trenton limestone which is exceedingly rich in fossils. The type locality, at Trenton Falls, is justly famous as a collecting place for Ordovicic fossils. Among plants, none above very simple seaweeds or algae are known to have existed. Among animals, hundreds of species have been described as occurring in the Ordovicic strata of New York. These species represent all the more important subkingdoms and classes of animals below the vertebrates. Especially prominent are: corals, graptolites, star-fishes, brachiopods, gastropods and trilobites. All the organisms mentioned lived in the salt water, and if land life forms existed we know practically nothing about them. It must be borne in mind that not a single species of that time is known to live today, so complete have been the evolutionary changes since the Ordovicic age. Certain remarkable classes of animals like the graptolites and trilobites, which often fairly swarmed in the Ordovicic sea, have been wholly extinct for millions of years.

## TACONIC MOUNTAIN REVOLUTION (CLOSE OF THE ORDOVICIC)

We are now ready to discuss the second well-known mountain-making epoch which affected New York State. We have learned that sedimentation along the middle eastern border and south-eastern parts of the State was practically uninterrupted during all the Cambric and Ordovicic periods, and that some thousands of feet of strata had accumulated. At the same time extensive sedimentation was taking place in the seas which covered all the regions of the present Berkshire hills, Green and White mountains, as well as southward, at least as far as Virginia, over the region occupied by the present Piedmont plateau. *At or toward the close of the Ordovicic period a great compressive force in the earth's crust was brought to bear upon the mass of sediments which reached from north of New England to Virginia, or possibly farther southward. As a result of this compression the strata were tilted, highly folded, and elevated far above sea level into a magnificent mountain range known as the Taconic mountains.* In structure, the range consisted of a series of rock folds, both great and small, whose axes were parallel to the main axis of the range, that is north-northeast by south-southwest. Examination of figures 14 and 20 will give the

reader a good conception of the character of the folding. It is quite the rule throughout this region of Taconic disturbance to find the strata either on edge or making high angles with the plane of the horizon. Many times the folds were actually overturned, and at times notable thrust faults or fractures[1] were developed, that is, the strata sometimes broke across and one great mass was pushed over another, as is well shown in many places in southeastern New York. These facts all go to indicate that the mountain-making compressive force applied to the region was of the extreme type, and though we have no way of telling just how high the range may have been, nevertheless the structural features and the vast amount of erosion since the folds were produced clearly indicate that the uplift was at least several thousand feet. The Green mountains, White mountains, Berkshire hills, Highlands-of-the-Hudson, and the Piedmont plateau are in a sense remnants of the great Taconic range.

In passing westward from the main axis of the Taconic range, the folding becomes less and less intense and finally dies out altogether. This fact is well illustrated by figures 6, 8 and 9. Along the Hudson river near Albany, the strata are fairly well folded, while a few miles westward the folds disappear. Passing eastward from Albany into Rensselaer county, one enters a region of excessive folding. In passing westward from Poughkeepsie, the intensity of the folding diminishes somewhat, but the shale formation is distinctly folded where it passes under the main mass of the Catskill mountains. Figure 6 clearly illustrates this fact.

How do we know that the Taconic disturbance occurred toward the close of the Ordovicic period? Another inspection of figure 6 will show that the strata of the next succeeding period (Siluric) rest directly upon the eroded edges of the folds of late Ordovicic rocks (see plate 25). Hence it is obvious that the disturbance occurred before the Siluric strata were deposited. What was the condition of the rest of the State just after the Taconic disturbance? In central New York, near Utica, a distinct eroded surface at the summit of the Ordovician shales proves that region to have been dry land toward the end of the period. On the north side of the Adirondacks and in the Champlain valley no formation younger than Ordovicic shale occurs, and all evidence points to uplift of that area into dry land toward the close of the period. Data are not obtainable for western New York. To summarize: *Practically all of northern-central, eastern, and southeastern New York (in-*

---

[1] See figures 23 and 24 for explanation of faults.

*cluding the great Taconic mountains) was dry land toward the close of the Ordovicic period, while the physical geography of western New York for that time is not certainly known.*

It should be noted in passing that the rocks of the Highlands-of-the-Hudson were, for a second time, clearly involved in mountain-making disturbances. The structural features of the Taconic mountains are finely exhibited in southeastern New York from Poughkeepsie to New York City, where one literally passes across the roots of the former great range. The distinct northeast-southwest trend of the topographic relief in this part of the State is due to the fact that the relief is still largely controlled by the Taconic folds and faults. The Hudson river has cut a deep channel across these structure lines, and along its banks excellent opportunity is afforded for the study of the rocks, folds, faults etc.

Another feature which must not be overlooked is the profound metamorphism of the strata along the main axis of the range. The very intense compression, under very high moist heat, caused the deeply buried strata along the main axis of uplift to become rather plastic, and hence the sediments became more or less foliated and crystallized into the various metamorphic rock types, the limestone becoming marble, the shale becoming slate or schist, and the sandstone becoming quartzite. Thus we have extensive marble quarries in southern Vermont, the slate in the quarries of Washington county, New York, and the Berkshire schist in the Berkshire hills of Massachusetts. In passing down the Hudson river from Kingston to New York City, the several stages in the metamorphism of the Ordovicic slate formation are finely illustrated. Thus, from Kingston to near Poughkeepsie the strata are distinctly folded but not metamorphosed; from Poughkeepsie to the Highlands, the strata are highly folded and partially metamorphosed, the shale layers nearly always having been changed to slate, while the associated and more resistant sandstone layers have escaped change; from the Highlands to New York City the rocks have been highly folded and metamorphosed, both shale and sandstone having been converted into schist locally called the Manhattan schist. For example, the rocks exposed in Central Park are Manhattan schists which are believed to have been originally Hudson River shales and sandstones which have become thoroughly crystallized by intense metamorphism.[1]

[1] Professor Berkey has recently suggested the possibility that the Manhattan schist may be Precambric in age; if so, the latter part of this statement does not apply.

Accompanying the Taconic disturbance and possibly aiding the metamorphism were minor molten rock intrusions in the form of dikes. These dikes break through late Ordovicic strata and hence can not be older than late Ordovicic. A fine example of one of these dikes on Manhattan island is shown in plate 24.

The great compressive force which folded and upraised the Taconic mountains did not accomplish its work suddenly. The force was slowly and irresistibly applied and the mass of strata was gradually bulged and bent, or fractured if near the surface, the amount of time required for the whole operation being perhaps very long but beyond estimate. Such a length of time is, however, so short compared with all known geologic history that we are accustomed to refer to the formation of such a mountain range as simply an event of earth history.

From these statements we see that, even before the range had attained its maximum height above sea level, a very considerable amount of erosion must have taken place. When the very first fold appeared above the ocean level, erosion began its work and continued with increasing vigor as the mountain masses got higher and higher. Thus we have the warfare between two great natural processes — the building up and the tearing down. So long as the building up process predominated, the mountain range increased in elevation, and we say the range was in its period of youth. When the opposing forces were about equally balanced, the range tended to remain at a constant elevation and we say the mountains were in the period of maturity. When the tearing down (erosive) process was predominant, we speak of the range as having been in old age. When the mountains have been completely worn down close to sea level (peneplain) we speak of the death of the range.

Here is an example of one of the remarkable procedures of nature. After millions of years of work by the deposition of thousands of feet of strata, layer upon layer on an ocean bottom, a great compressive force is brought to bear and a magnificent mountain range is literally born out of the ocean. No sooner is this great mountain range well formed than the destructive processes unceasingly destroy this marvelous work. But the sediments derived from the wear of this range are carried into the nearest ocean again to accumulate and after long ages to be raised up into another range; and so the process is often repeated. From this we learn that the mountain ranges of the earth are by no means all of the same age. The Adirondacks are older than the Taconics, and these older than the Appalachians; the latter, in turn,

being older than the Sierras and the Coast ranges. Much of the material making up the mass of the Catskill mountains was derived from the wear of the Taconic mountains, deposited in the sea just to their west and later raised high above sea level.

## SILURIC PERIOD

The close of the Ordovicic age or the opening of the Siluric found practically all the State above sea level and undergoing erosion. Along the eastern side the great Taconic range stood out prominently, but over the rest of the State we have no evidence that the land was very high. The central portion of the Adirondack region probably stood out somewhat more prominently than the western region.

As shown on the geologic map (figure 1), the Siluric strata outcrops in a comparatively narrow belt which runs along the western side of the Hudson valley to the Helderberg hills, southwest of Albany, where it swings sharply around westward to follow the south side of the Mohawk valley, and thence as a somewhat wider belt along the south side of Lake Ontario. These Siluric strata everywhere dip under the Devonic (surface) rocks of the Catskill and Southwestern plateau provinces. This fact, combined with the knowledge that the strata are largely of widespread marine origin and also outcrop abundantly in central Pennsylvania, makes it practically certain that the Siluric rocks underlie all of the Catskill and Southwestern plateau regions. *Thus we must conclude that at least during much of Siluric time all of New York State south of Lake Ontario and the Mohawk valley and west of the Hudson river, was covered by sea water.* That the earliest Siluric sea did not spread over the area is proved by the absence of the very earliest known Siluric deposits. Furthermore, we know that the sea transgressed upon the State from the south or west, the Taconic range forming an effective barrier on the east and total absence of Siluric strata in the St Lawrence and Champlain valleys (as well as in Canada just north of the State) precluding encroachment of the sea from the north.

This encroachment of the sea over so much of the State was due to a gradual sinking of the land. That central and western New York was submerged before the Hudson valley region is proved as follows: In central New York (south of Utica) the first deposit to form upon the eroded surface of the Ordovicic shales was the Oneida conglomerate which passes westward into the Medina sand-

stone. In southeastern New York (for example, the Shawangunk mountain) the first Siluric deposit to be laid down upon the eroded Ordovicic shales was the Shawangunk conglomerate. This latter formation, as determined by its fossils, belongs with the Salina division and is therefore much younger than the Oneida conglomerate which belongs with the Medina division (see table in chapter 1). Also the whole of the Clinton and Niagara formations, which are so well developed in central and western New York, were never formed in eastern or southeastern New York. Thus the Siluric sea, due to subsidence of the land, overspread central and western New York long before it reached the Hudson valley region. In fact it was not until late in the period that the sea encroached upon the Hudson valley area, and then it did not occupy all that area because the shore of the Siluric sea extended only as far east as the western slope of the Taconics. Western New York, during the late Siluric, was a more or less cut off basin or arm of the sea in which the salt beds were deposited.

How much, if any, of the Adirondack region was covered by the Siluric sea? The total absence of any formation later than the Ordovicic shales around the northern Adirondacks and across the line in Canada strongly suggests that this region was upraised toward the close of the Ordovicic period, perhaps at the same time as the Taconic revolution, and continued as dry land not only during the Siluric but also during all the ages up to the present, except for a very brief local submergence during the Quaternary (see figure 34). In the southern Adirondack area the case is somewhat different. The outcrops of Siluric strata beneath the steep front of the Devonic Helderberg escarpment immediately south of the Adirondacks, makes it certain that these strata and the Siluric sea formerly extended farther north. The difficulty comes in trying to decide how far northward these rocks once extended, because there is now not a single scrap of Siluric rocks north of the Mohawk river, though the cap rock (Oswego sandstone) of the Tug Hill plateau is probably of Siluric age. All we can say is that the Silurian sea probably overspread the southern border of the Adirondacks and that the sediments which were deposited there have since been removed by erosion. To summarize: *During the early Siluric the sea had spread over only central and western New York, while during the late Siluric it had extended over practically all the State west and south of the Adirondack region.*

The strata of Siluric age were deposited sheet upon sheet in the usual manner upon the sea bottom. For our purpose we may con-

sider that sedimentation was uninterrupted, though as a matter of fact there were certainly minor oscillations of level which inter-

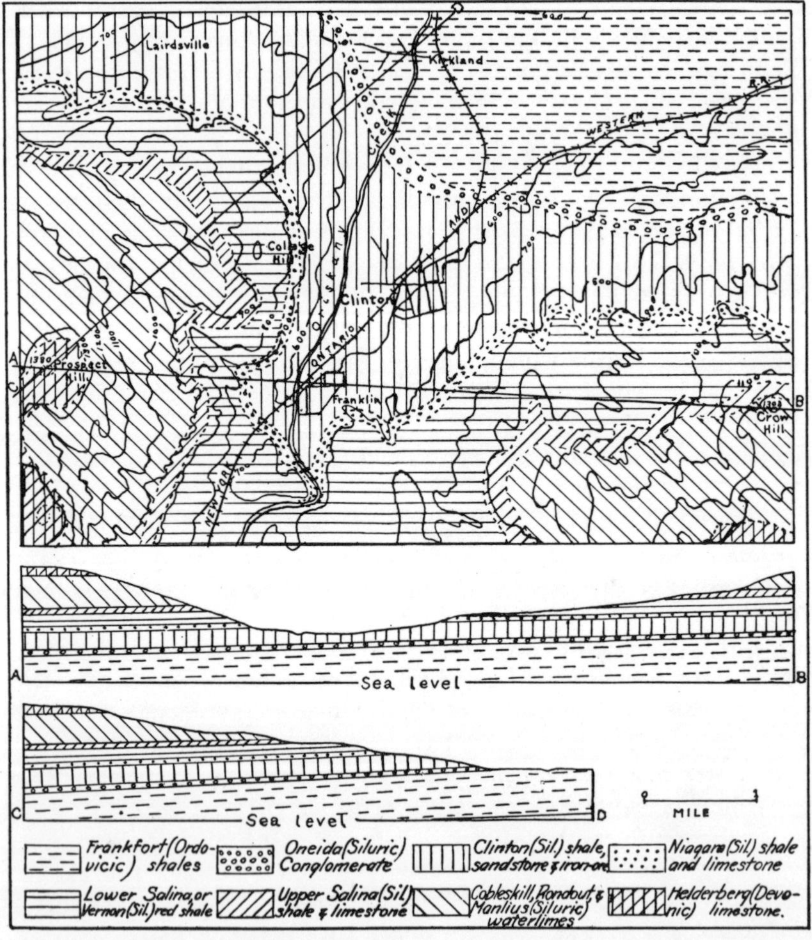


FIG. 17 Geologic and topographic map and structure sections of the vicinity of Clinton (Oneida county) showing the surface distribution and underground relations of the various rock formations from the Upper Ordovicic to the Lower Devonic inclusive. Note the simple nonfolded and nonfaulted structure, and the gentle southwesterly dip (tilt) of the formations. A similar simple structure characterizes the formations of the whole southwestern plateau province. Vertical scale of the sections four times exaggerated.

Geology by W. J. Miller

fered with the deposition of sediments and produced slight unconformities. These minor interruptions have not yet been carefully

studied and hence need not be considered here. The total thickness of Siluric strata along the line of outcrop in the State varies considerably. In central and western New York the thickness is generally from 1000 to 1500 feet, while in southeastern New York it is much less since only the thinned upper formations are present.

The first Siluric sediments to form in central and western New York are called the Oneida conglomerate and Medina sandstone, these two being of practically the same age. These coarse deposits were washed into the shallow sea from the northern lands, that is in Canada and the Adirondack region. Next in order came the deposits of Clinton age, which consist of layers of shale, sandstone and iron ore. Above the Clinton come the Lockport and Guelph formations made up of shales and dolomitic limestone, the limestone forming the crest of Niagara Falls. None of the formations, so far mentioned, extend to the Hudson valley, but with the opening of the great Salina epoch Siluric deposits for the first time reached to the Hudson valley region where the earliest rock to form was the Shawangunk conglomerate which rests upon the eroded Ordovicic shales at the summit of Shawangunk mountain. This rock is entirely confined to southeastern New York, while rocks of the same age in central and western New York are shales and limestones. In this latter region the lowermost (oldest) Salina formation is the Vernon red shale, usually from 100 to 300 feet thick, which, in the western part of the State, is overlain by the salt and gypsum beds. Deep wells have proved the presence of the salt beds under practically all the Southwestern plateau. The salt was deposited in great salt lagoons and the climate of the time must have been arid. With an influx of fresh water into the lagoons, the type of deposit changed, and the hydraulic limestone (water lime) beds were formed all the way across the State to the Hudson valley region. These water lime beds are quarried at many places along the line of outcrop across the State, but more especially in the famous Rosendale cement region (see plate 33). Next in order, and marking the summit of the Siluric, come the Cobleskill, Rondout, and Manlius limestones which, though not very thick, are remarkably persistent across the State.

As to the life of the Siluric seas it may be said that it is in effect the continued existence of the same organic groups that preceded in the waters of Ordovicic time, though some diminished, some increased and some new ones made their first appearance. Thus the graptolites and trilobites greatly diminished, while the echinoderms (star fishes) increased, and the brachiopods and mollusks held

their own. Fossil seaweeds, but not animal remains, are common in the Oneida-Medina beds. Various fossils exist in profusion in the Clinton and Niagara formations, while in the middle Salina beds fossils are altogether absent because the water of that time was intensely saline. The waterlime beds at the base and top of the Salina are usually poor in fossils except for the remarkable assemblage of organic remains known as eurypterids which belonged to the arachnid class but are now wholly extinct. Fossils are generally rather common in the uppermost Siluric beds of the State.

## DEVONIC AND CARBONIC PERIODS

The Devonic history of New York State is comparatively simple and the records are remarkably well shown in rocks of that age. *Devonic strata comprise the whole Catskill and Southwestern plateau provinces, except for a few small patches of Carbonic rocks, and thus cover more than one-third of the area of the State. They are more widespread on the surface than the rocks of any other age.* The combined thickness of the Devonic strata is over 4000 feet, which is considerably more than for any other Paleozoic period in the State.

That the Devonic strata, on the Hudson valley side, formerly extended some miles farther eastward than they now do is proved by the presence of small outliers of Devonic rock, for example, Becraft mountain just southeast of Hudson, the Rensselaer grit farther north and the Skunnemunk mountain southwest of Newburgh. During part of the time the Devonic sea, or arms of it, reached as far east as these outlying masses and doubtless far beyond over the regions of Massachusetts and the Connecticut valley. The bold outcropping edges of thick Devonic strata facing the Mohawk valley and, in the Helderberg escarpment, the Ontario plain, make it certain that the strata formerly extended some distance farther northward. It is more than likely that this northward extension of Devonic rocks was not beyond the southern border of the Adirondacks; at least we have no positive knowledge that the Devonic sea ever covered any of northern New York (see figure 18).

There was no disturbance of any kind at the close of the Siluric, so that period passed very quietly into the Devonic. The Oriskany sandstone was for many years regarded as the base of the Devonic but now, as a result of a careful study of the fossils, the line between Siluric and Devonic is drawn just below the Helderberg lime-

stone. As is the case with the Siluric, the rock formations are piled one upon another like great sheets and all show a gentle southward dip (see figures 3 and 5). Beginning at the bottom, the Helderberg limestone was succeeded in regular order by the Oriskany sandstone, Onondaga limestone, Marcellus and Hamilton

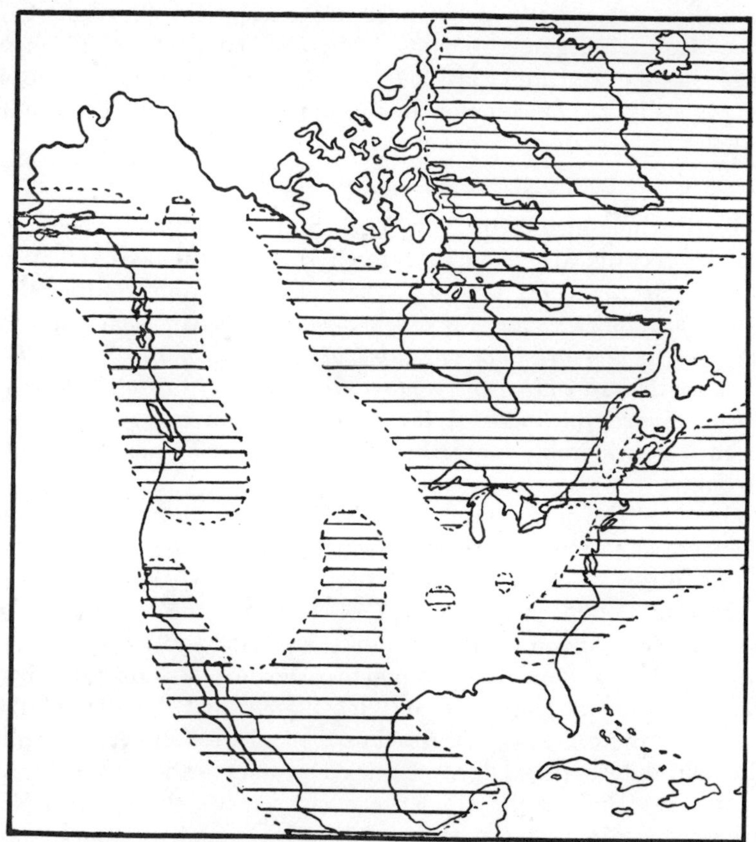

FIG. 18 Generalized map of North America in the Devonic period, showing the relations of land and water. Horizontal lined areas = land; blank areas = water. Only the northern and extreme southeastern portions of the New York State area were land. The western shore of Appalachia was farther west than during the Ordovicic, due to the addition of the Taconic mountain area

shales, all of which were deposited over the whole Devonic basin in New York. Above these come the Tully limestone and Genesee shale which extend from east-central to western New York. Still higher, and forming the summit of the Devonic, are the Portage shales and sandstones and the Chemung (or Catskill) sandstones

which extend from the Catskill mountains to western New York.

Except for the comparatively thin Tully formation, the limestone is confined to the Lower Devonic and the lower part is not more than a few hundred feet thick. Thus the great bulk of Devonic rock lies above this limestone and consists of shales and sandstones piled layer upon layer. These latter rocks are clearly land-derived sediments which were washed into the Devonic sea by streams from the Taconics and also probably from land areas which are known to have existed to the north in Canada.

The Devonic strata, from oldest to youngest, abound in the fossils of marine organisms, and some fossil land plants have also been found. Looked upon in a broad way, Devonic life was much like that of the Siluric, though certain fundamental differences are to be noted. Thus the Devonic furnishes the first really authentic evidence of the existence of land plants. Such plants as ferns, lycopods (club mosses), and equisetae (horse tails) grew to be large treelike forms and in considerable profusion. Remains of these have been found in the Devonic strata in New York. All of them belonged to the very simple, nonflowering plants and were closely related to the plants of the next succeeding Carbonic (coal) period. Among the fossil animals especially abundant in the Devonic rocks of the State are: sponges, corals, echinoderms (star fishes), brachiopods, mollusks (including the bivalves, gastropods, and cephalopods), and arthropods (including trilobites and eurypterids). The graptolites became extinct during the early Devonic. One of the remarkable features of the life was the great abundance and variety of fishes, so that this period is commonly referred to as the "Age of Fishes." From the zoological standpoint all the fishes were of simple types, the true bony skeletons of modern fishes being entirely absent. Devonic fish remains in considerable numbers have been found in the State.

Carbonic strata are only very sparingly represented in New York, there being a few small outlying masses in the southwestern portion of the State (Cattaraugus and Allegany counties). Immediately southward, in Pennsylvania, Carbonic strata are developed on a great scale so we can be certain that the Carbonic sea spread over the southern border of New York State. It is quite possible that this sea extended over most of southern New York, but positive evidence, due to absence of strata, is lacking.

The Permic is the last great period of the Paleozoic era, but rocks of that age are nowhere present in New York State.

## APPALACHIAN MOUNTAIN REVOLUTION (CLOSE OF THE PALEOZOIC)

The Paleozoic era was brought to a close by one of the most profound physical disturbances in the history of North America. It has been called the Appalachian revolution because at this time the Appalachian mountain range was born out of the sea by upheaval and folding of the strata. Because of the direct effect of this great upheaval upon the history of New York State, a brief description is given.

All through the vast time (probably ten million years) of the Paleozoic era, a great land mass existed along what is now the eastern coast of North America. This land, which has been called Appalachia, had its western boundary approximately along the present coast line, while it must have extended eastward at least as far as the present border of the continental shelf. Concerning the altitude and character of the topography of Appalachia we know almost nothing, but we do know that it consisted of metamorphic rock of Precambric age, and very similar to that of the Adirondacks. The tremendous amount of derived sediments shows that Appalachia was high enough during nearly all its history to undergo vigorous erosion. Although oscillations of level more than likely affected Appalachia, and its western shore line was quite certainly somewhat shifted at various times, nevertheless it persisted as a great land-mass with approximately the same position during all its long history. Its general position is well shown on the map, figure 16.

Barring certain minor oscillations of level, all the region just west of Appalachia was occupied by sea water during the whole Paleozoic era, and sediments derived from the erosion of Appalachia were laid down layer upon layer upon that sea bottom. The coarsest and greatest thickness of sediments was deposited nearest the land, that is along what we might call the marginal sea bottom. At the same time finer sediments, in thinner sheets, were being deposited all over the Mississippi valley region. By actual measurement, in the present Appalachians, we know that the maximum thickness of these sediments was at least 25,000 feet. These are all of comparatively shallow water origin, as proved by the coarseness of sediment, ripple marks, fossil coral reefs etc., and so we are forced to conclude that this marginal sea bottom gradually sank during the process of sedimentation, thus producing what is called a great geosynclinal trough. Perhaps the very weight of accumu-

lating sediments caused this sinking. Finally, toward the close of the Paleozoic era, sinking of the marginal sea bottom and deposition of sediments ceased, and "eventually the trough began to yield to lateral compression and its contained strata were thrown into folds or fractured by great overthrusts. Thus in place of a sinking sea bottom along the shore of the great interior sea, arose the Appalachian mountains, which in their youth may have been a very lofty range rivalling the Alps in height. This range extends from the mouth of the St Lawrence river to Alabama."[1] As a result of this great physical revolution practically all of eastern North America was raised well above sea level, though the more moderately elevated Mississippi valley region was unaccompanied by folding or faulting of the strata.

*The effect of the Appalachian revolution upon New York State is of fundamental importance because the whole State, except probably a small area near the mouth of the present Hudson river, was raised well above the sea, and true marine conditions never again prevailed over any part of its area except the extreme southeastern portion.*[2] Judging by the vast amount of erosion which took place during the succeeding Mesozoic era, we are safe in our belief that the general elevation of the State at the close of the Paleozoic was at least several thousand feet above sea level. It is also important to note that this great uplift in New York was accomplished without any folding of the strata except along the Hudson valley. The gentle southward to southwestward tilt (dip) of the Paleozoic strata, however, is thought to have been produced at this time due to somewhat greater uplift on the north.

Along the western side of the Hudson valley, folds produced at the time of the Appalachian revolution are plainly visible, though the folding of the rocks here was much less violent than in the Appalachians proper. As a matter of fact these folds are but continuations of those of eastern Pennsylvania, but the compressive force in southeastern New York was too weak to cause much disturbance. Professor Davis has aptly styled these, "Little mountains east of the Catskills." By far the most conspicuous physiographic feature of this folded region is the Shawangunk mountain (ridge) which stands out very prominently and whose

[1] Scott's Introduction to Geology, second edition, p. 647.

[2] The influx of tide waters along the eastern and northern borders of the State in the Quaternary period presents no exception to this statement because the conditions then were esturaine rather than marine.

very existence is due to the fact that, as a result of the folding and subsequent erosion, the great sheet of hard and resistant conglomerate has been left as a protective cap over the soft Hudson river (Ordovicic) shales (see figure 19). In the Rosendale cement region the effects of the folding are also evident (see figure 10 and plate 33). The folds in the Siluric and Devonic strata of Skunnemunk mountain were also produced at this time. Of course the whole lower Hudson valley was subjected to this mild compressive force but, since all the rocks older than the Siluric were already so greatly disturbed, it is often impossible to see the effects of the Appalachian disturbance. Thus we see that mountain-building forces have affected the rocks of the Highlands-of-the-Hudson at least three times (Precambric, Taconic revolution, and Appalachian revolution); Cambric and Ordovicic strata of the lower Hudson valley twice (Taconic and Appalachian revolutions); and the Siluric and Devonic strata but once (Appalachian revolution).

The extensive faulting or fracturing of the eastern Adirondack and Mohawk valley regions is a matter of no small importance in our discussion of the physical history of the State, because the present major topographic features of those regions are largely dependent upon the faulting. It is generally believed that much of this faulting occurred toward the close of the Paleozoic era, and most likely at the time of the Appalachian revolution, but since considerable faulting certainly occurred later than that time, it is thought best to discuss this whole subject toward the close of the next chapter.

## Chapter 5

# MESOZOIC HISTORY

### TRIASSIC PERIOD

We have observed that as a result of the Appalachian revolution New York State was raised well above sea level and this was its condition at the opening of the Triassic period. The total absence of any Triassic strata of marine origin makes it quite certain that the continent extended farther eastward than it does today and if so, the old Paleozoic land mass called Appalachia still existed, though probably much diminished in height by this time. The absence of marine rocks, however, does not mean that no deposition of Triassic sediments occurred within the borders of the State, because a remarkable series of nonmarine strata which were accumulated along the Atlantic slope are, in part, shown in southeastern New York (Rockland county and Staten island).

These nonmarine strata are of Upper Triassic age, as told by the fossils, and their present distribution and mode of occurrence clearly show that they were deposited in a series of long troughlike depressions whose trend was parallel to that of the main axis of the Appalachian range. These troughs lay between the Appalachians proper and old Appalachia. The latter also was now partly made up of the greatly worn-down Taconics. The facts that these troughs are truly downwarps, and that they so perfectly follow the trend of the Appalachian folds, make it certain that they were formed by a great lateral pressure which was a continuation of the Appalachian disturbance. Thus the Appalachian mountains still seem to have been growing well into the Triassic period, and while the Paleozoic strata were being folded the surface of old Appalachia, including part of the Taconic region, was also more or less warped, and the downwarps formed the troughs in which the Triassic beds were deposited. One of these troughs extends along the Connecticut river through Connecticut and Massachusetts; another, and the largest, reaches from Rockland county, New York, through northern New Jersey, southeastern Pennsylvania, Maryland, and into northern Virginia; while several smaller ones lie in Virginia and North Carolina. These depressions were most favorably situated for rapid accumulation of thick deposits because of their position immediately between the two great land masses which

were being eroded. The sediments derived from the erosion of the young Appalachians were especially abundant because of the vigorous wearing down of the young mountains. A thickness of thousands of feet of nonmarine rocks, mostly red sandstones and shales, was finally accumulated in these basins, and is known as the Newark series. The great thickness of these rocks, from 10,000 to even possibly 15,000 feet, strongly argues for a gradual downwarping of the basins as deposition of sediments went on. It is often stated that these strata were formed in estuaries, but at least in the northern area, from the Connecticut valley to Maryland, many of the layers show sun cracks, rain-drop pits, ripple marks, and remains and footprints of land reptiles. These features show that for the most part the beds were formed in very shallow water such

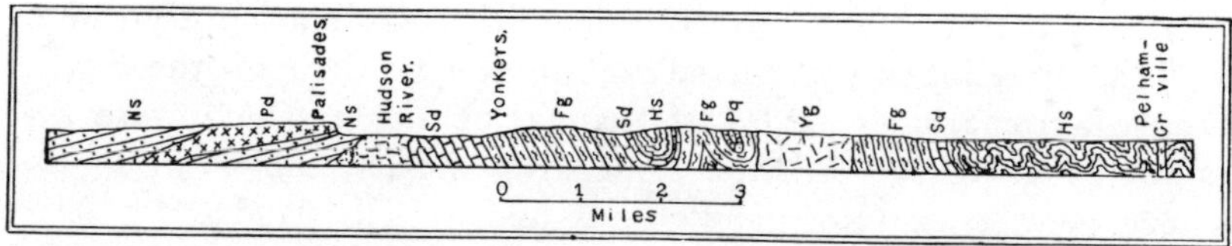


FIG. 20 Detailed section running west-northwest through Pelhamville, Yonkers and the Palisades in southeastern New York, showing the intense folding of Taconic age, the granite dikes, and the relation of the Palisade lava to the other formations. Fg = Fordham fineiss (Precambric); Pq = Poughquag quartzite (Cambric); Sd = Stockbridge dolomite (Cambro-Ordovicic); Hs = Hudson schist (Ordovicic); Yg = Yonkers gneiss; Gr = granite dike; Ns = Newark sandstone (Jura-Trias); Pd = Palisade diabase or lava (Triassic).

Modified from New York City folio, U. S. G. S.

as flood plains or lakes where changing conditions frequently allowed the surface layers to lie exposed to the sun.

During the time of the formation of the Newark beds there was considerable igneous activity as shown by the occurrence of sheets of igneous rocks within the mass of sediments. In some cases true lava flows, with cindery tops, were poured out on the surface and then became buried under later sediments, while in other cases the sheets of molten rock were forced up either between the strata or obliquely through them, thus proving their intrusive character. As a result of subsequent erosion, these lava intrusions often stand out conspicuously as topographic features. Perhaps the most noteworthy of these is the great igneous rock sheet, part of which outcrops to form the famous Palisades of the Hudson and which outcrops altogether for a distance of seventy miles. As shown in

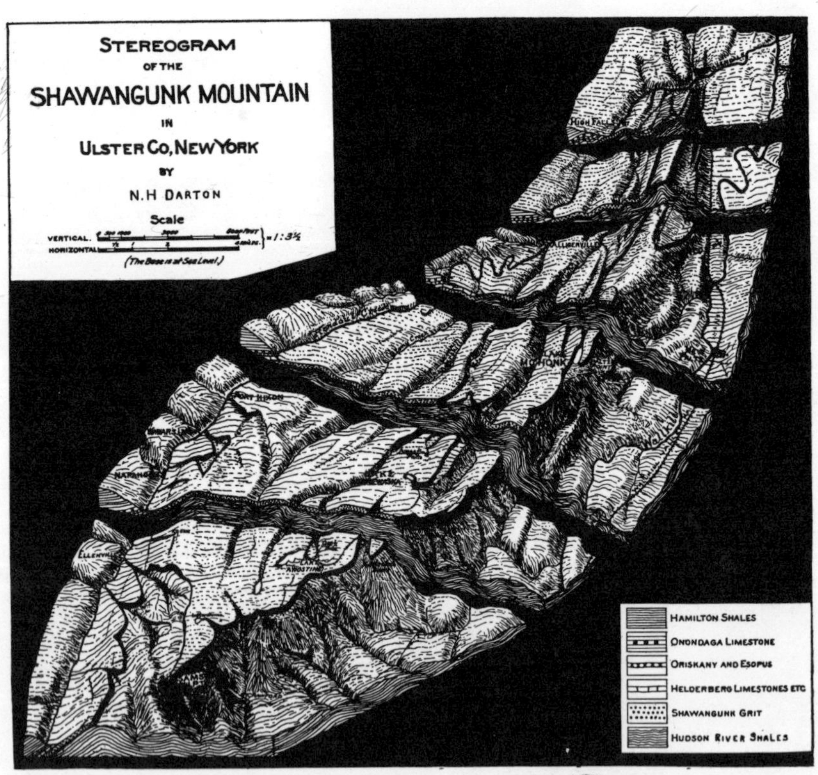


FIG. 19 The main body of the mountain consists of a great thickness of folded Ordovicic (Hudson River) shales and sandstones which are capped by a comparatively thin, but very resistant layer of Shawangunk conglomerate.

After Darton N. Y. State Mus. Rep't 47, 1894, facingp. 540

figure 20, the molten rock sheet first broke through the strata and then crowded its way along parallel to them. During the process of cooling there was contraction which expressed itself by breaking the rock mass into great, crude, vertical columns, and hence the origin of the name "Palisades" (see plate 34). At the base of the Palisade rock, as well as on its top a little back from the edge of the cliff, the Newark sandstone outcrops. The steep cliff is due to the fact that the hard igneous rock is much more resistant to erosion and weathering than the sandstone above and below it.

The rocks of the Newark series are nearly everywhere somewhat folded, tilted and extensively fractured by normal faults. Just when this deformation occurred is not exactly known, but it was probably at the close of the Triassic period as will be shown under the next heading.

Briefly summarized, *the Triassic was a time of accumulation of thick deposits of red sandstone and shale of nonmarine character in troughlike depressions along the Atlantic slope, these deposits being represented in southeastern New York. During their formation there was considerable igneous activity when sheets of lava were forced through or between the strata as is well shown in the case of the rock of the Palisades.*

## JURASSIC PERIOD

No rocks of Jurassic age occur within New York State nor as a matter of fact in all eastern North America, except possibly some fresh-water deposits along the Potomac river of Maryland. The failure of such strata is readily explained by the fact that the Jurassic period was ushered in by a slight upwarping (accompanied by faulting and tilting of the rocks) of the Atlantic border of North America so that there were no basins of deposition within the present eastern border of the continent. That this uplift actually occurred and that the Jurassic period in the eastern United States was a time of extensive erosion, is well established because the whole Atlantic seaboard, including the tilted and faulted Triassic strata, was worn down well toward the condition of a peneplain and the next sediments (Cretacic) were deposited upon the eastern portion of that worn-down surface (see figure 22). For instance, on Staten island and in northern New Jersey, the Cretacic beds may be seen resting directly upon the deeply eroded Triassic rocks, and hence the proof is conclusive that during much if not all of the Jurassic period active erosion was taking place, and this in turn implies that the Triassic beds were well elevated in the early Jurassic.

Briefly summarized: *No deposition of Jurassic strata occurred within the borders of New York State but instead, the region was well above water and undergoing active erosion so that by the close of the period this region, as well as the whole Atlantic slope, had been worn down to the condition of a fairly good peneplain.*

## CRETACIC HISTORY

The Cretacic period opened with the eastern coast line of the eastern United States somewhat farther out than it now is, but early in that period there was enough subsidence, or possibly warp-

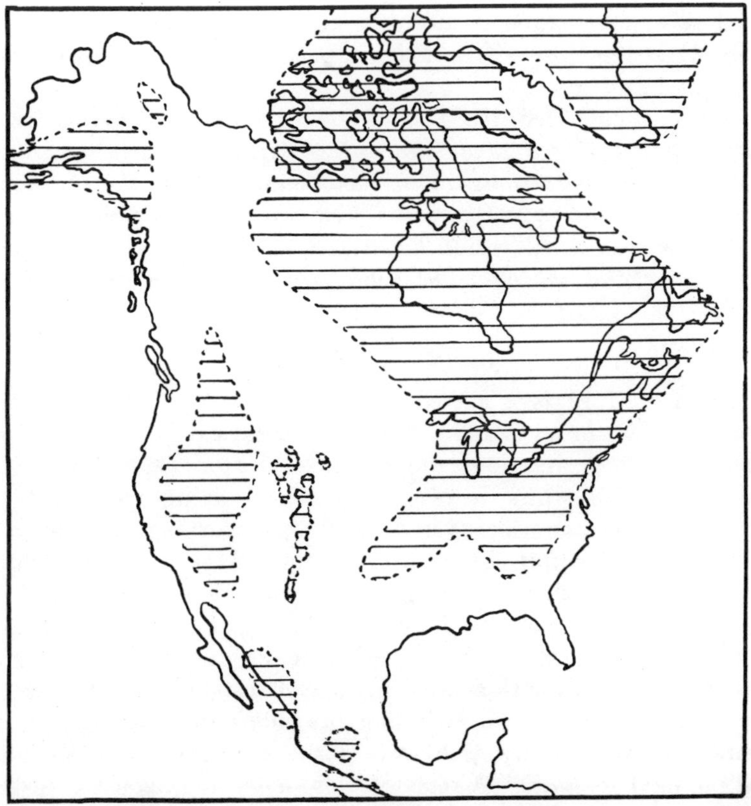

FIG. 21 Generalized map of North America in the Upper Cretacic period, showing the relations of land and water. Horizontal lined areas = land; blank areas = water. The sea then spread over the Atlantic coastal plain region including Long and Staten Islands of New York. During the next (Teritary) period, the conditions were much the same along the Atlantic and Gulf coasts, but in the west the great interior sea had disappeared.

ing, of the coastal lands to allow deposition of sediments over much of the region now known as the Atlantic Coastal plain. That but little downwarping of the surface was necessary in order to produce proper conditions for sedimentation is evident because the coastal lands, just prior to the Cretacic, were already low-lying as a result of the long Jurassic erosion period. There was just enough warping of these low coastal lands to produce wide flats, flood plains, shallow lakes, and marshes back from the real coast line and in which were deposited the sediments derived from the Piedmont plateau and Appalachian areas. The early Cretacic deposits thus formed are known as the Potomac series, and consist of very irregular layers of sand, gravel and clay. The very irregular arrangement of these beds and their rich content of fossil land plants, afford conclusive evidence that the sediments were not accumulated under marine conditions. The Potomac series outcrops at the western margin of the present Coastal plain and has been traced from Martha's Vineyard, through Nantucket, Long Island, Staten island, Northern New Jersey, and southward into Georgia. Passing seaward the strata dip under those of later age (see figure 22). On Long Island, Potomac outcrops occur only along the northwestern border but these beds no doubt dip under the more recent deposits of the rest of the island. The maximum thickness of the Potomac series is only about 700 feet.

Along the Atlantic coast certain deposits which should come between the Lower and Upper Cretacic are missing, and the Upper Cretacic beds rest upon the eroded surface of the otherwise undisturbed Lower Cretacic. Thus we know that there was a gentle upward oscillation of the land toward the end of the Lower, or beginning of the Upper, Cretacic, after which a moderate amount of erosion of the Lower Cretacic beds took place.

Then came another gentle submergence of the coastal lands when the Upper Cretacic strata were formed. The character and present extent of these deposits, and the fact that they are of marine origin, prove that this subsidence allowed a shallow sea to spread over practically all of what is now called the Atlantic Coastal plain including most of Long and Staten islands in New York. *Accordingly we learn that, for the first time since the close of the Paleozoic, did truly marine conditions prevail over any portion of New York State, and also that Appalachia, the great land mass of the east, which had persisted through the many million years of the Paleozoic and most of the Mesozoic, now disappeared under the Cretacic sea.*

The present surface distribution of the Upper Cretacic beds is much like that of the Lower Cretacic, and they also dip under the still later formations of the Coastal plain (see figure 22). The thickness of the Upper Cretacic is never more than a few hundred feet.

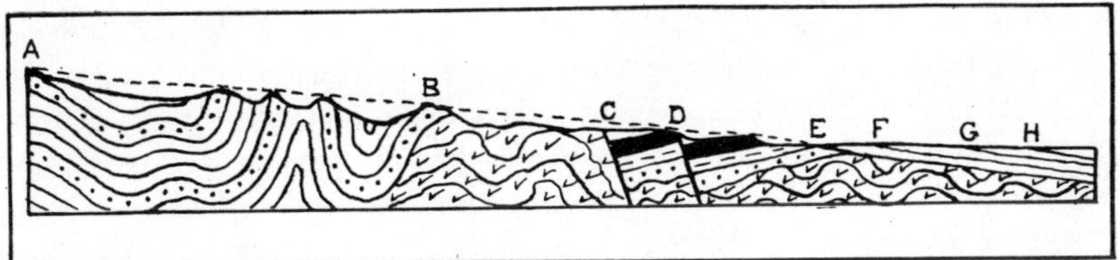


FIG. 22 Diagrammatic section through the Atlantic slope, at about the latitude of northern New Jersey, showing the structures and relations of the various physiographic provinces as they now exist.

A to B = Folded Paleozoic strata of the Appalachian mountains, with hard strata standing out to form the ridges.

B to C = Piedmont plateau consisting of highly folded and metamorphosed rocks of Precambrian and early Paleozoic ages.

C to E = Triassic strata showing tilting and faulting of the beds and mode of occurrence of an igneous rock sheet (D) which outcrops to form a low ridge.

E to H = Coastal plain, consisting of comparatively thin sheets of unconsolidated sediments.

E to F = Cretacic beds (upper and lower).

F to G = Tertiary beds.

G to H = Quaternary beds

H = Present coast line.

The dotted line represents the peneplain character of the surface (except for the tilting) at the close of the Cretacic period.

To summarize: *The Cretacic period opened with slight subsidence of the Coastal plain region, including southeastern New York, to produce low-lying flats upon which the nonmarine Potomac sediments were deposited. Then came a slight reelevation (accompanied by erosion) followed by subsidence of the Coastal plain region enough to allow encroachment of the shallow sea in which the Upper Cretaceous sediments were accumulated.*

## LIFE OF THE MESOZOIC

The life of the Mesozoic is but scantily represented within New York State because rocks of that age are so poorly exposed. The Mesozoic era is commonly referred to as the "Age of Reptiles" because animals of that class then reached their culmination of

development. During this era the great dinosaur reptiles, the largest land animals that every lived, stalked the western plains. Some remains of smaller dinosaurs have been found in the Triassic beds of the Atlantic coast, one specimen lately having been discovered in the Newark beds along the lower Hudson. Reptilian tracks abound in the Newark strata. Mammals appeared in the early Mesozoic, but throughout the era they continued small and comparatively insignificant. The first birds and true bony fishes (teleosts) appeared in the later Mesozoic, but they are either absent or not important in the Mesozoic of the middle Atlantic coast. The invertebrate life of the era was in general very different from that of the Paleozoic, few types from the latter era having persisted, and by the end of the Mesozoic the invertebrates took on a decidedly modern aspect.

Among plants, those of the early era were still simple nonflowering kinds much like those of the Carbonic or "Coal age," while in the late Mesozoic flowering plants of very modern types, including many of our present forest trees, were prominently developed. The Cretacic beds of the Atlantic coast are rich in fossil plants.

## THE CRETACIC PENEPLAIN AND ITS UPLIFT

During all the Mesozoic era most of the eastern portion of the United States was above water and undergoing erosion, so that, as a result of this very long period of wear, the region was reduced to the condition of a more or less perfect peneplain. It is known as the Cretacic peneplain because of its best development during the Cretacic period. This vast plain extended over the areas of the Appalachian mountains, Piedmont plateau, all New York State, the Berkshire hills, and the Green mountains. Its most perfect development was in the northern Appalachians, for example, from east-central Pennsylvania to Virginia, where hard and soft rocks alike had been so thoroughly cut down that no masses projected notably above the level of the low-lying plain.

Farther northward, however, over New York and western New England, its development was less perfect so that certain masses of harder rock stood out more or less prominently above the general level of the plain. In the central and eastern Adirondacks many low mountains of very resistant igneous rock rose above the peneplain surface. In a similar manner an occasional low mountain stood out in the Berkshire Hills region, and it seems probable that the hard Devonic sandstones of the Catskills also rose notably above the peneplain, though in the latter case positive proof has not been

given. Thus, *toward the end of the Mesozoic era all the area of New York State had been reduced to a vast, monotonous, featureless plain (peneplain) except for the mountain masses of very moderate elevation in the east-central Adirondack, and possibly also the Catskill, regions.* As Professor Berkey says: "The continent stood much lower than now. Portions that are now mountain tops and the crests of ridges were then constituent parts of the rock floor of the peneplain not much above sea level. This rock floor was probably thickly covered with alluvial deposits (flood plain) not very different in character from the alluvial matter of portions of the lower Mississippi valley of today. Upon such a surface the principal rivers of that time flowed, sluggishly meandering over alluvial sands and taking their courses toward the sea (the Atlantic) in large part free from influence by the underlying rock structure. The ridges and valleys, the hills, mountains and gorges of the present were not in existence, except potentially in the hidden differences of hardness of rock structure. Such conditions prevailed over a very large region, certainly all of the eastern portion of the United States."[1]

In the western part of the United States the Mesozoic era was brought to a close by what must take rank as one of the greatest mountain upheavals in the history of North America. This is known as the Rocky Mountain revolution because the great Rocky Mountain system was chiefly formed at this time. At the same time *in the eastern part of the United States the Mesozoic was closed by an important physical disturbance though on a far less grand scale than that of the west. This disturbance produced an upwarp of the vast Cretacic peneplain with maximum uplift of from two to three thousand feet following the general trend of the Appalachians and thence through northern New York.* This upward movement was unaccompanied by any renewed folding of the strata, and the effect was to produce a broad dome sloping eastward and westward, and northward to the Gulf of St Lawrence and southward to the Gulf of Mexico.

A prominent effect of this great uplift was to revive the activity of the streams so that they once more became active agents of erosion. *We are now prepared to make the important statement that the present major topographic features of New York State, as well as western New England and the whole Appalachian region, have largely been produced by the erosion or dissection of this up-*

[1] N. Y. State Mus. Bul. 146, p. 67.

*raised Cretacic peneplain.* This being the case, are any remnants of that upraised surface still left? In the affirmative answer to this inquiry we have the most positive evidence for the former existence of the Cretacic peneplain. We have said that the most perfect development of the peneplain was from central Pennsylvania to Virginia, and it is just here where we should expect to find the best remnants of that old surface. In this region the typical Appalachian ridges and valleys, which run parallel to the trend of the mountain range, are very well developed. These valleys are the trenches cut along the belts of soft rock and to below the surface of the upraised peneplain, while the ridges have developed along the belts of hard rock and their summits actually represent portions of the old peneplain surface. These ridges all rise to the same general level for miles around, and as viewed from the summit of any one of them the concordant altitudes give rise to what is called the "even sky line" which is a most striking feature of the landscape. Plate 35 gives an excellent idea of the even sky line across these ridges.

In New York State the concordant altitudes are not so well shown both because the peneplain was here not so perfectly developed and because the attitude of the strata has largely been unfavorable to the formation of long, distinct ridges. Remnants of the peneplain are, however, unmistakably present in New York as, for example, on a very large scale over the great Southwestern plateau whose high points nearly always rise to altitudes of about 2000 feet. This plateau is simply a part of the upraised and dissected Cretacic peneplain, and the slight downward sag toward the middle (already noted in chapter 2) is no doubt due to a slight downwarping of the general level during the process of uplift. The topographic map (plate 4) well illustrates the character of this dissected plateau. The present elevation of the peneplain remnants does not necessarily indicate the maximum amount of uplift. In the next chapter evidence will be presented to show that, for the New York area at least, the land was considerably higher in the Tertiary period than it is at present.

The summit of the Tug Hill province is a small plateau at an altitude of about 2000 feet, and is merely a remnant of the upraised peneplain which was formerly connected with the Southwestern plateau. As one stands at the summit of Tug hill and looks out over the western slope of the Adirondacks, he is impressed by the remarkably even sky line there shown at an altitude of a little over 2000 feet. The east-central Adirondacks and the Catskills present exceptions because these regions stood out above the old peneplain.

The Mohawk and upper Hudson valleys have been so broadly and deeply trenched through soft strata that in them no remnants of the peneplain surface remain. Immediately eastward in the Berkshires, however, the old surface is well exhibited. In the Highlands-of-the-Hudson, a view from one of the high points shows a rather even sky line at an altitude of from 1200 to 1500 feet, the somewhat lower elevation of the old surface here being due to the fact that this region was east of the main axis of uplift.

Since the actual work of erosion or dissection of the upraised peneplain occurred during the Cenozoic era, further discussion of the subject is reserved for the next chapter. It has been the present purpose to prove that the Cretacic peneplain actually existed and that it was upraised.

## FAULTING OF THE EASTERN ADIRONDACKS

*The eastern and southern Adirondack regions have been extensively fractured or faulted* (see figures *23 and 24*). *In fact the major topographic features of those regions such as the numerous north-northeast by south-southwest ridges and valleys are largely*

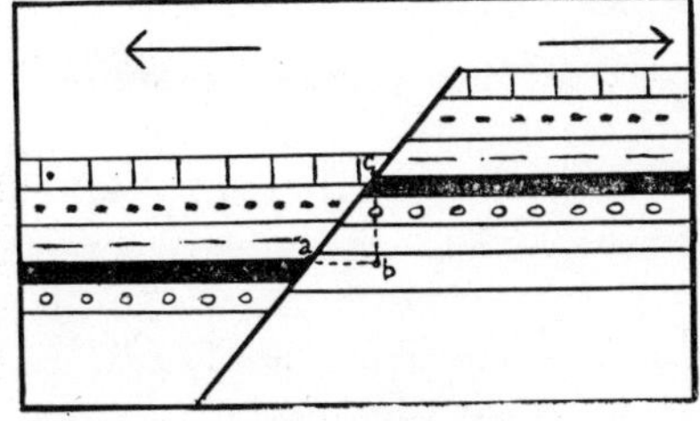

FIG. 23 Cross-section of a normal fault.

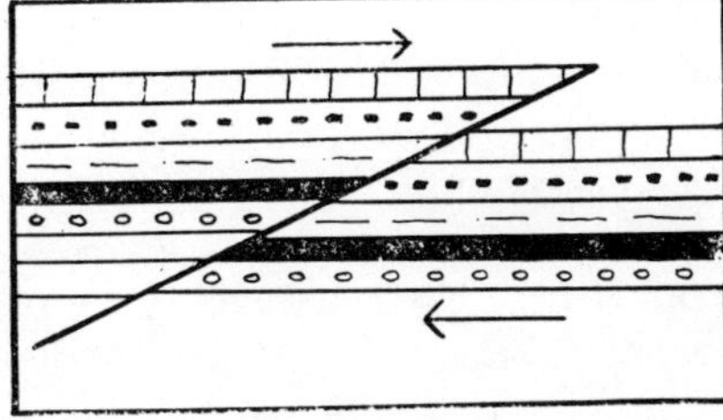

FIG. 24 Cross-section of a reversed or thrust fault.

*dependent upon this faulted structure.* These fractures are all of the normal fault type with fault surfaces practically vertical. Examination of the topographic maps of the whole eastern and southern Adirondacks shows that by far most of the ridges and valleys, streams and lakes trend in a north-northeast by south-southwest direction, or perfectly parallel to the direction of the major faults. Up to the present, no single fault has been proved to extend across the entire region, but rather there is a series of numerous parallel faults, no one of which has been traced much over 20 or 30 miles. The exact amount of displacement along these lines of fracture in the ancient crystalline rocks can not be determined, but many times it amounts to at least 2000 feet.

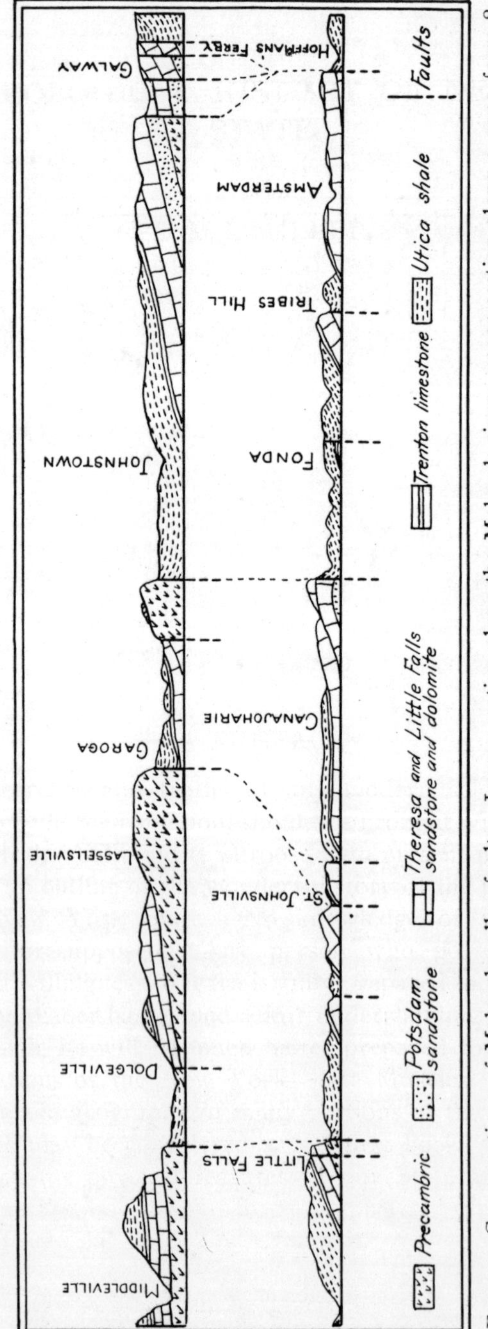


FIG. 25 Cross-sections of Mohawk valley faults. Lower section along the Mohawk river; upper section along a zone from 8 to 10 miles farther north. Length of each section 52 miles; vertical scale greatly exaggerated.
Modified after Darton, Annual Rep't N. Y. State Geol., 1894

This series of faults cuts through the early Paleozoic strata along the shores of Lake Champlain and in the Mohawk valley, and in these regions, due to marked differences in the rocks affected, it has been possible to determine carefully the character of the faults and the amounts of the displacements (see plate 37). Figure 25 shows two sections through the faulted region of the Mohawk

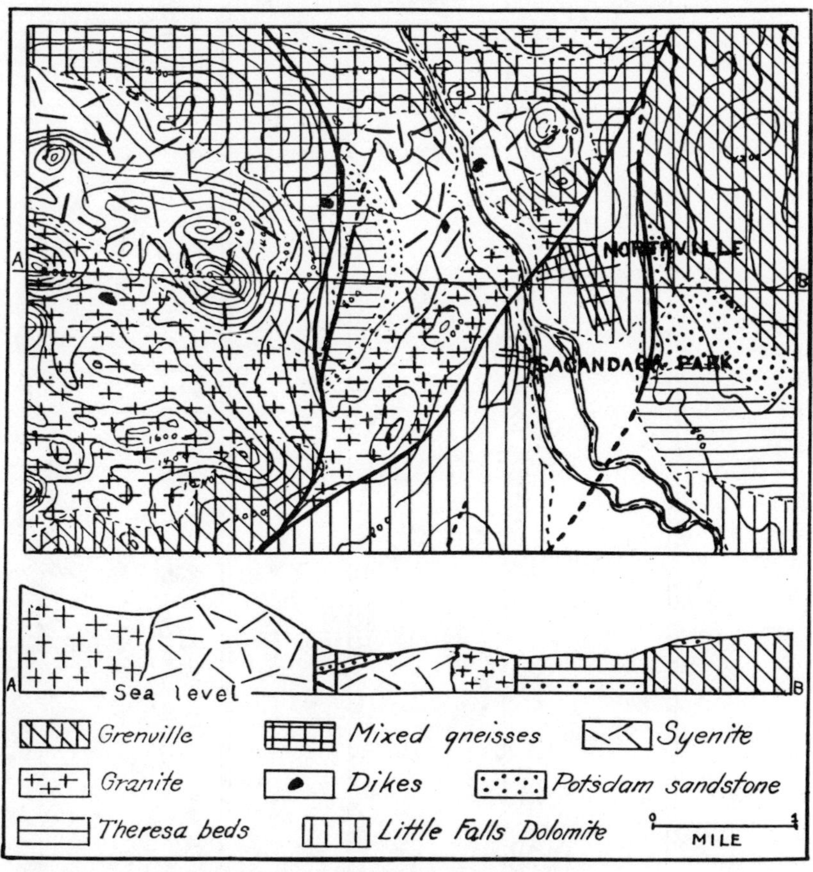


FIG. 26 Geologic and topographic map of the vicinity of Northville (Fulton county), showing an unusual variety of rock formations and structures along the southern border of the Adirondacks where the Precambric and Paleozoic rocks come together, and where all have been greatly faulted. The position of the structure section is indicated by the line AB, the vertical scale of the section being twice exaggerated. The greatest fault is the one on the west, and the country immediately on the east side of it has dropped fully 1500 feet with respect to that on the west (mountain) side. Northville lies between two smaller faults and on an earth block which has dropped several hundred feet with respect to the country on either side.

Geology by W. J. Miller, N. Y. State Mus. Bul. 153

valley. To a considerable degree the topography of the valley is affected by the faults, especially where the harder Precambric or Cambric rocks form the scarp or upthrow sides of the faults. This is particularly true at Little Falls and the "Noses" (near Yosts) where, in each case, the Mohawk river has cut a gorge across a prominent fault scarp and even down to the underlying Precambric rock which has been brought relatively nearer the surface by the tilting of the earth blocks (see figure 7). On the geologic map of the State (figure 1) two tongues of Paleozoic rock are seen to extend northward well into the Precambric rock area, and these are to be explained by the fact that, due to faulting along the west sides, the Paleozoic strata, for fifteen or twenty miles, have dropped down (relatively) fully 1500 feet with respect to the Precambric rock. The much more resistant Precambric rock has stood out against erosion and in each case rises with steep front from 1000 to 1500 feet above the Paleozoic rock surface (see figure 26). The small remnant of Paleozoic strata already referred to at Wells in Hamilton county was dropped down fully 2000 feet by faulting against the Precambric rock just west, and thus this remarkable Paleozoic outlier has been preserved from complete removal by erosion.

What is the age of the faulting or, in other words, when were these fractures developed? That some faulting, at least, occurred during Precambric time has been well established but, so far as known, those faults are of very minor importance, certainly having no appreciable influence upon the existing topography.

During the Paleozoic era, however, there is good reason to think that considerable faulting took place. At just what time during the era the faulting occurred is not now altogether certain, but it is certain that it was sometime after the deposition of the Ordovicic sediments because at many places those rocks are involved in the faulting. Cushing has suggested that the faulting may have been initiated at the time of the Taconic revolution when the rocks of the region immediately eastward were so greatly disturbed, but he says, "the great earth disturbances (Appalachian revolution) which prevailed in the Appalachian zone toward the close of the Paleozoic would seem more likely to have brought about the major faulting of the reigon."[1] At this latter time the rocks of northern New York were not folded but, as we have learned, the whole State was notably elevated, and during this disturbance conditions were cer-

[1] N. Y. State Mus. Bul. 95, p. 405.

tainly favorable for extensive fracturing of the strata. The disturbance of the early Mesozoic, which caused the fracturing of the Newark (Triassic) rocks along the Atlantic slope, quite certainly did not affect the Adirondacks because those faults are of a different type and closely confined to the Triassic basins.

If the major faulting occurred at the close of the Paleozoic, then the Mesozoic must have opened with the northeastern portion of the newly upraised New York State area cut by a great series of faults which caused the edges of the upturned earth blocks to stand out prominently as ridges. However this may have been, we are certain that by the close of the long period of Mesozoic erosion the old fault scarps or ridges were practically obliterated. If so, how do we account for the present Adirondack ridges which follow the fault lines? As a result of the uplift of the Cretacic peneplain one or both of the following things happened, either there was renewed faulting, or that as a result of unequal erosion (due to differences in rock character) on opposite sides of the faults, the old fault scarps were renewed. It is quite certain that both things occurred, and thus the surface of the newly elevated Cretacic peneplain in northeastern New York was made irregular by freshly formed fault scarps. This, together with the later (Cenozoic) erosion along the old fault lines and belts of weaker rocks, accounts for the existing Adirondack ridges. That some of the faulting actually dates from this time, or possibly even later, is proved by the present existence of certain steep fault cliffs in perfectly homogeneous rock masses, and by the fault blocks which have been scarcely modified by erosion since their formation.

## DRAINAGE OF NEW YORK IN THE MESOZOIC

Thus far we have said very little about the early drainage features of New York State. In fact we must admit that prior to the Cenozoic era, we have practically no knowledge concerning the positions of even the major drainage lines of the State. From our knowledge of the land and water relations during the Paleozoic era, we can form only the most general ideas regarding the drainage. The Mesozoic physiography of the State is better known and hence the drainage is perhaps better known, but even here positive knowledge is almost wholly lacking. The whole subject of the Precenozoic drainage of New York is one which demands thorough study before anything like satisfactory conclusions can be reached, and the following very brief discussion is intended to be merely suggestive of the problems involved.

At the close of the Paleozoic, and as a result of the Appalachian uplift, the region of New York State was raised well above sea level, with the greatest uplift toward the north as shown by the general south to southwesterly dip (tilt) of the Paleozoic strata (see figures 3 and 5). At that time those strata lapped over much of what is now the Precambric rock area of the Adirondacks. The Appalachian folds of the Hudson valley region, as well as the highlands (of earlier origin) in general along the eastern border of the State, must have prevented any important eastward drainage. Thus, in the writer's belief, the strongest evidence suggests that the principal streams of the early Mesozoic era flowed in general southwesterly courses upon the surface of the newly upraised Paleozoic strata and away from the highlands of the eastern border of the State.[1]

If this be the correct interpretation (for others are certainly possible) of the early Mesozoic drainage, it must follow that no river at all comparable in length and position to the present Hudson could have existed along the eastern side of the State, and no large rivers, like the Susquehanna and Delaware, then had southeasterly courses across the Appalachian mountains.

During the long Mesozoic era, the area of the State was profoundly eroded, as already proved. In the midst of this era the ruggedness of Mesozoic relief reached its maximum and, in accordance with well-known principles, the valleys must have formed along the belts of softer rock, while the harder rocks stood out to form the highlands or ridges. At this time the edges of the Paleozoic strata had sufficiently retreated (by erosion) on all sides from the central Adirondacks so that a considerable area of Precambric rocks had already become exposed in northern New York. During this retreat of the Paleozoic strata there was a tendency to form important valleys, especially along the western and southern borders of the Adirondacks, because whenever the harder rock formations were encountered they would stand out as escarpments, while the softer rocks would be worn down into valleys. It is in accordance with these principles that the Mohawk and Black river valleys were formed, though it does not necessarily follow that these older valleys occupied the same positions as the present ones because of the gradual retreat of these depressions away from the Adirondack region.

---

[1] It should be stated that the Great Lakes were not then in existence, those bodies of water not having been formed till late in the Cenozoic era (see chapter 6).

In western New York and over the region of the present Lake Ontario, the hard and soft early Paleozoic strata outcropped along a nearly east and west direction, and hence considerable streams, tributary to the major southwestward flowing streams, doubtless followed the belts of soft (shale) rocks. Such a west-flowing stream may have followed the belt of weak Ordovicic shales which runs under the present Lake Ontario.

In southeastern New York, in the midst of the Mesozoic era, the land was lower than at the beginning of the era as shown by the fact that the late Cretacic sea spread over at least some of the region. This gave a better opportunity for the development of an eastward or southward drainage toward the Atlantic basin, and at this time it is possible that the ancestors of the modern Hudson, Delaware, and Susquehanna rivers were formed.

However uncertain our ideas may be regarding the topography and drainage of the early and middle Mesozoic, we are nevertheless sure that by the close of the period the topography of the State was that of almost a peneplain which has already been described, and that the streams were all of low gradient and very sluggish. During the long erosion time of the Mesozoic, there must have been many changes in stream courses and adjustments to rock structures. By the close of the era the courses of the rivers are, as yet, not definitely known, though in accordance with the above discussion we are reasonably certain that the principal drainage of the State from the northern, central, and western portions was southwestward to westward into the Mississippi basin, while the drainage of the southeastern portion was southward to southeastward into the Atlantic basin.

## Chapter 6

# CENOZOIC HISTORY

## TERTIARY PERIOD

**Rock formations and life of the Tertiary** The Mesozoic closed and the Cenozoic opened with the uplift of the great Cretacic peneplain. Before the uplift, the sea spread over the Long and Staten Islands region, but for a time after the uplift the land was there high enough to exclude the sea and New York State was wholly above sea level. This we know because the lowest (earliest) Tertiary deposits do not occur on Long or Staten Islands or in northern New Jersey, and hence the region must have been above water. The subdivisions of the Tertiary, from oldest to youngest, are known as Eocene, Miocene, and Pliocene. The early Eocene deposits are missing from the northern Atlantic Coastal plain, and on Long and Staten Islands we have no evidence that any of the Eocene is present which thus leads to the conclusion that all southeastern New York was dry land during the whole of the Eocene. During the Miocene there was enough sinking to allow the sea to encroach over the Long and Staten Islands districts as well as the whole northern Coastal plain. Except for very slight oscillations of level which we shall here disregard, the region remained submerged under shallow sea water during all the Miocene and Pliocene, or till the close of the Tertiary period. The Tertiary deposits were sands, gravels, and clays which formed layer upon layer in the shallow sea along the margin of the continent (see figure 22), but on Long and Staten Islands they are seldom seen because of the more recent covering of glacial deposits. They are finely exposed in the Coastal plain of New Jersey.

The Tertiary period is generally called the "Age of Mammals" because, although mammals began in a small way in the Mesozoic, they became the dominant feature of life for the first time in the Tertiary. In the early Tertiary the mammals were very different in appearance from those of the present, a common form then being a generalized or ancestral type (for example, Phenacodus) about the size of a dog and having five toes. Many of our modern mammals have descended from this type. During the Tertiary the mammals developed very rapidly so that by the close of the period they were very much as they are today except that man, the highest

and most wonderful animal of all, did not appear until the last (Quaternary) period of earth history. Birds developed to much like their present forms. Reptiles diminished both in size and number of species in a remarkable way, while fishes took on a decidedly modern aspect. The invertebrates of the Tertiary were not strikingly different from those of the present and by the close of the period they were so modern that from 75 to 90 per cent of them were even the same species as those now living.

We learned that, in the late Mesozoic, true flowering plants had been developed in abundance, and during the Tertiary these and all other plants reached a development which in no essential way was different from that of the present.

The records of Tertiary life are but scantily represented in New York because of the small extent of exposed Tertiary rocks. In the deposits of the Atlantic Coastal plain, however, abundant fossils are found.

**Development of relief features.** The uplift of the great Cretacic peneplain was an event of prime importance for New York because it literally furnishes us with the beginning of the history of most of the existing relief features of the State. Hence we assert with emphasis that all the principal topographic features of the State as we see them today date from the uplift of the Cretacic peneplain because they have been produced by the dissection of that upraised surface. This dissection was largely the work of erosion, though in the eastern Adirondack region faulting has produced notable effects. All the great valleys such as the Champlain, St Lawrence, Black river, Mohawk, and Hudson have been produced since the uplift of the peneplain. It should also be stated that the Great Lakes, as well as the numerous lakes, gorges, and waterfalls for which New York is noted, were absent as geographic features at the opening of the Cenozoic.

As previously stated, the streams of New York which flowed upon the peneplain surface sluggishly meandered over deep alluvial or flood-plain deposits, and their courses were little if any determined by the character of the underlying rocks because hard and soft rocks alike were worn down to a general level. The uplift of the peneplain, however, greatly revived the activity of the streams so that they became very effective agents of erosion; they first cut channels through the alluvial deposits and then into the underlying bedrock. Thus these large original streams had their courses determined in the overlying deposits, and when the underlying rocks were reached the same courses had to be pursued

entirely without reference to the underlying rock character and structures. Such streams are said to be superimposed because they have, so to speak, been let down upon the underlying rock masses. To quote Professor Berkey: "The larger rivers, the great master streams, of the superimposed drainage system, in some cases were so efficient in the corrasion of their channels that the discovery of discordant structures (in the underlying rocks) has not been of sufficient influence to displace them, or reverse them, or even to shift them very far from their original direct course to the sea. They cut directly across mountain ridges because they flowed over the plain out of which these ridges have been carved and because their own erosive and transporting power have exceeded those of any of their tributaries or neighbors."[1] Fine examples of such superimposed streams which are now entirely out of harmony with the structure of the regions over which they flow are the Susquehanna, Delaware, and Hudson. Thus the Susquehanna cuts across a whole succession of Appalachian ridges while, in accordance with the same explanation, the Delaware cuts through the Kittatinny range at the famous Delaware Water Gap. The lower Hudson pursues a course no less out of harmony with the structure of the country through which it passes. Thus it flows at a considerable angle across the Taconic folds above the Highlands, after which it passes through a deep gorge which it has cut through the hard granites and other rocks of the Highlands. The simple explanation is that the Hudson had its course determined upon the surface of the upraised Cretacic peneplain, and that it has been able to keep that course in spite of the discordant structures of the underlying rocks.

But while the great master streams were thus cutting deep trenches in hard and soft rock alike, numerous side streams or tributaries came into existence and naturally developed along the belts of weak rock and in harmony with the geologic structures. This is true of all the streams now occupying the valleys between the Appalachian ridges. In southeastern New York two remarkable cases are presented by the Wallkill river and Rondout creek which flow many miles northeastwardly and in a direction almost the reverse of that of the Hudson to which they are tributary. As the master superimposed Hudson cut its channel deeper and deeper, the Wallkill and Rondout side streams were enabled to cut their valleys deeper and deeper while they increased in length by pushing

[1] N. Y. State Mus. Bul. 146, p. 69.

their headwaters farther southward, but only along belts of weak rock and in harmony with the northeast-southwest folded structure of the region.

In a similar manner the great Mohawk valley has been developed. The Mohawk is the chief tributary of the Hudson, and whether or not its ancestor flowed in about the same position upon the surface of the peneplain, we do know that the present Mohawk valley (below Little Falls) has been carved out of the upraised peneplain by the Mohawk river and its tributaries along a belt of weak Ordovicic shales. The valley is bounded on the north by the very hard Precambric rocks and on the south by the fairly resistant limestones of the Helderberg escarpment. Proof will be given on a later page for the statement that late in the Tertiary (that is just prior to the great Ice age) the Mohawk river had its source near Little Falls, and that at the same time another stream (Rome river), now extinct, flowed westward from Little Falls, past Utica and Rome and into the basin now occupied by Lake Ontario. West Canada creek was then tributary to the Rome river, while the Sacandaga river flowed into the Mohawk instead of the Hudson as it now does.

However uncertain we may be as to the location of a Precenozoic Susquehanna river, we are very certain that the present numerous, deep channels of the Susquehanna drainage system have been cut into the upraised peneplain. The Susquehanna, like the Hudson, is a good example of a superimposed stream and its ancestor may have flowed along the same general course over the low-lying peneplain before its uplift. Immediately after the uplift some of the more easterly headwaters of the Susquehanna came out of the southern Adirondacks. Evidences of this are as follows: (1) The present Mohawk valley had not been formed, the Mohawk river then having only begun the westward migration of its headwaters along the belt of soft shales; (2) the natural slope of the southern Adirondack region was then southward into the east branches of the upper Susquehanna; and (3) the positions of the present sources of the east branches of the Susquehanna at the crest of the high Helderberg escarpment, and in some cases within a very few miles of the present Mohawk river (see drainage map, figure 11) strongly argue for the cutting off or beheading of the former headwaters of the east branches. This beheading was accomplished by the Mohawk as its headwaters migrated slowly westward thus tapping one by one of the upper waters of the east branches of the Susquehanna. Even today the Mohawk continues to steal drainage

territory at the expense of the Susquehanna because the short, swift tributaries of the Mohawk, which flow northward over the Helderberg escarpment, are cutting down their channels rapidly, while their headwaters are migrating southward into the territory of the Susquehanna. A great network of large and small streams tributary to the upper Susquehanna drain a considerable portion of south-central New York, there being no single great master stream in this region because the rock formations are so nearly horizontal and are so much alike as regards resistance to erosion.

The Delaware system has had a history practically the same as that of the Susquehanna, except that it never drained any of the region north of the Mohawk.

The present ruggedness of the Catskills is largely, if not altogether, due to the production of deep channels which have been cut into the region upraised at the time of the uplift of the great peneplain by the headwaters of the Delaware, Schoharie creek (north-flowing), and the smaller streams flowing across the steep eastern front of the mountains.

During Tertiary times Lake Champlain was certainly not in existence, but the great depression was there and was no doubt largely developed or at least increased in depth by the settling of earth blocks during the time of extensive faulting at the close of the Mesozoic or beginning of the Cenozoic. The depression is essentially a fault trough. The major stream occupying this valley flowed northward and in late Tertiary time, at least, the divide between the drainage of this and the Hudson valley passed between Glens Falls and Whitehall, and through the present position of the " Narrows " of Lake George, the lake, of course, not then being in existence.

**Drainage of New York in the Tertiary.** The outline of the probable drainage condition of western New York during the Mesozoic has already been given, and now as we attempt to restore the drainage conditions of the Tertiary, we must admit that some problems yet remain unsolved. Lakes Ontario and Erie certainly were not in existence. Streams flowed through these basins, which were not as deep as they now are. The bottom of Ontario is as much as 491 feet below sea level, while its surface lies at an altitude of 247 feet, and the altitude of Erie is 573 feet while its greatest depth is 204 feet. The explanation of the increased depths of the basins is given on a later page. The question now arises, Did the waters from western New York drain westward or southwestward and into the Mississippi, or northeastward through the St

Lawrence? According to J. W. Spencer, the St Lawrence received those waters, but in the light of what we have said about Mesozoic drainage, and also in view of the fact that the St Lawrence now, in the Thousand Islands region, does not flow through anything like a distinct channel (see plate 9) cut out by a great river, we must admit that there is little to favor such northeastward drainage from western New York. The St Lawrence is almost certainly postglacial in its course at the Thousand Islands as shown by the lack of any real channel, and by the presence of a belt of hard Precambric rock extending across the river and connecting the Adirondacks with the Canadian Precambric rocks. This hard rock belt must have formed a preglacial divide until the recent formation of Lake Ontario and the downwarping of the land which allowed the drainage to pass over the divide for the first time (see later page).

Grabau's interpretation is that the Tertiary drainage of western New York passed westward and southwestward into the Mississippi, and this view is, in the writer's belief, far more tenable. The accompanying map (figure 27) gives a good idea of the drainage lines according to this view. The major drainage lines were no doubt inherited from the Mesozoic, the southwestward courses having been originally determined by the tilt of the land at the time of the Appalachian uplift. When the Cretacic peneplain was upraised, these major streams as, for example, the Dundas river, again began very active work of erosion, and tributary streams were developed, during the Tertiary, along the belts of weak rocks. Thus an important west-flowing tributary was developed along the belt of soft Ordovicic and Medina shales, and formed a channel where the basin of Lake Ontario now is. The Rome river, with source at Little Falls, became a branch of this stream while another important branch had its source on the Thousand Islands divide. There is also shown the position of the Black river, which by late Tertiary had already carved out that important valley and flowed into the Ontario depression, according to Grabau. More recent evidence, however (see later page), strongly favors the passage of the lower end of the Black river north and northeastward into the precursor of the modern St Lawrence, and which had its source on the Thousand Islands divide.

On the accompanying map the three south-flowing streams, one heading near Rochester and the other two flowing through Lakes Seneca and Cayuga, are, in the writer's judgment, not properly shown for the late Tertiary, that is, just prior to the great Ice age.

It is practically certain that preglacial streams were here north-flowing rather than south-flowing, because during long Tertiary time such tributaries to the large stream in the Ontario basin must have developed across the steep north slopes of the Niagara and Helderberg escarpments, and also because the slope of the Genesee

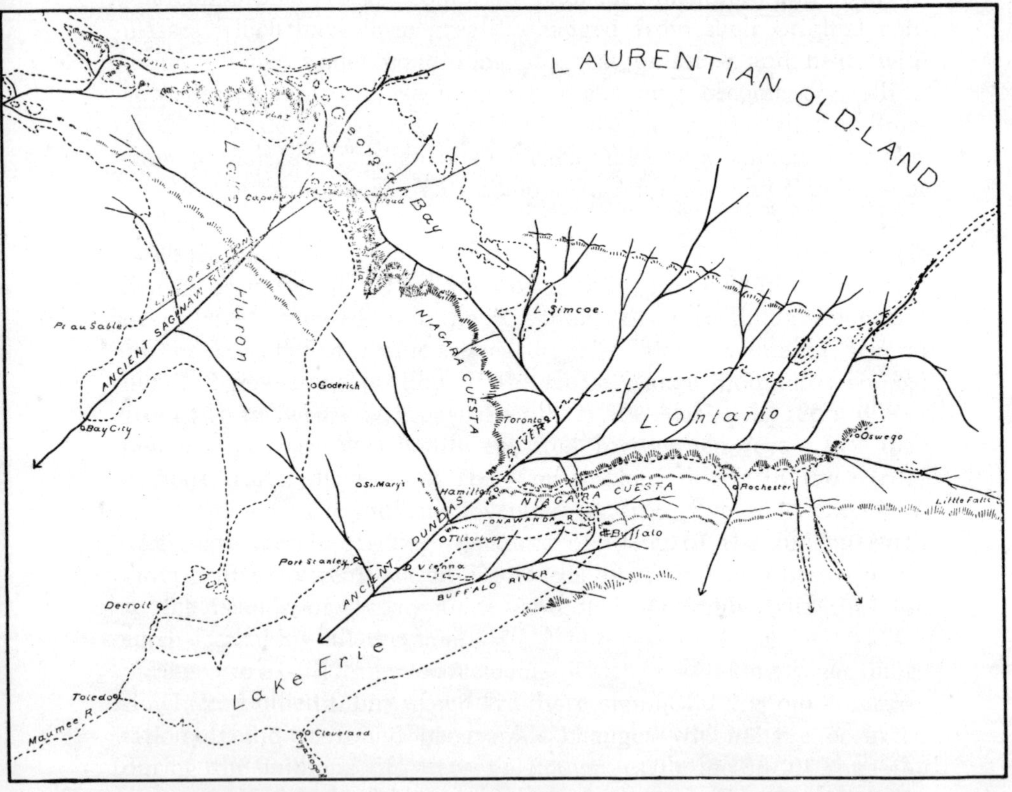


FIG. 27 Grabau's interpretation of late Tertiary drainage in the eastern Great Lakes region, the intention being to show the general kind of drainage rather than the exact location of the streams. The "Rome" river with source at Little Falls is well shown. The three south-flowing streams should, in the writer's opinion, be represented as north-flowing, at least immediately prior to the great Ice Age.

After Grabau, N. Y. State M s. B. l. 45, fig. 6

river is now, at least, so decidedly downward toward the north. If the late Tertiary Genesee flowed southward, the reversal of its course must have been caused by a very marked tilting of the land, but we have no evidence for such a decided land movement.

In our discussion of Mesozoic drainage (see chapter 5) we found that, during the early part of that era, nearly all the drainage of

the state passed southwestward to westward into the Mississippi valley with very little, if any, drainage into the Atlantic. But by the close of the Mesozoic, or the beginning of the Tertiary, a considerable south to southeastward drainage had been established along the lines of the Hudson, Susquehanna, and Delaware rivers. Have we any explanation for the establishment of these important drainages into the Atlantic? One view is that the courses came about as a result of the meandering and changing of channels on the surface of the low-lying Cretacic peneplain, and that these courses were maintained upon the upraised peneplain. Another view which the writer would suggest as worthy of consideration is that the south and southeasterly drainage lines were established as a result of the uplift of the peneplain. Chapter 5 shows that the axis of greatest uplift passed through northern New York from central Pennsylvania, the region of southeastern New York and northern New Jersey not being raised so high. It is very reasonable, if not highly probable, that this very warping or slow arching up of the peneplain surface inaugurated the present drainage toward the Atlantic because the streams, no matter what their previous courses, then naturally must have flowed down that initial slope toward the Atlantic.

After the uplift of the peneplain the larger streams cut down their channels most rapidly and would be the first to reach "grade," that is, a condition in which, because of low velocity, they could no longer cut down their channels, though the widening process could continue because of side cutting due to meandering of the streams back and forth from one side to the other of the channels. The deep but broad-bottomed, stream-cut valleys so common in New York State show that many of the streams had reached a graded or nearly graded condition even by the close of the Tertiary. In southeastern New York, at least, we have evidence to show that after the streams had reached grade there was an appreciable renewed uplift of the land which again revived the activity of the streams. Thus the broad Hudson valley, with minor hills rising above its surface, was produced when the Hudson was well along toward a graded condition and then, as a result of this late Tertiary uplift of the land, the present narrow and fairly deep inner channel (gorge) of the Hudson was formed. The Hudson did not reach grade in this inner channel, its work having been interrupted by the spreading of the great ice sheet over the region. It is not known

that this late Tertiary reelevation notably affected the rest of the State.

This inner gorge of the Hudson valley has been traced for fully 100 miles eastward beyond the mouth of the present river. The Coast and Geodetic Survey has made a detailed map (see figure 28) of the ocean bottom near New York City, and the submerged channel of the Hudson river is clearly shown as a distinct trench cut

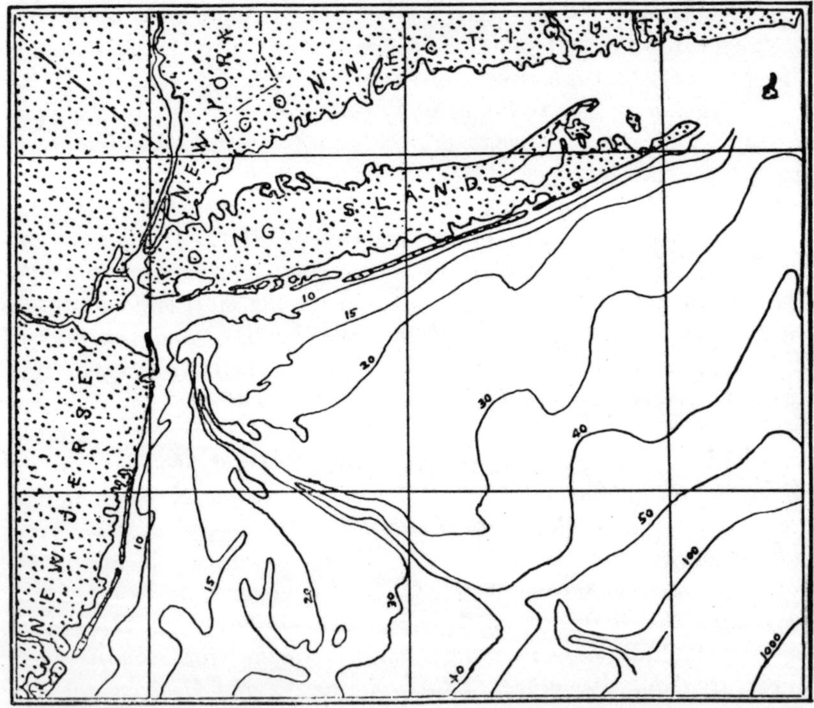


FIG. 28 The submerged Hudson River channel, whose position is clearly shown by the contour lines. Figures indicate depth of water in fathoms. Data from Coast and Geodetic Survey.

into the continental shelf. Even in the Hudson valley above New York City, the narrow inner rock channel has a depth of hundreds of feet (see plate 38) and is mostly submerged below tide water. Without any question this submerged Hudson channel was cut when the region was dry land, and thus we have positive proof that late in the Tertiary, and possibly extending into the early Quaternary, the region of southeastern New York was notably higher than it is today. Conservative estimates place the amount

of elevation greater then than now at not less than 2000 feet because the very end of the Hudson channel is submerged to that extent.[1] The coast was then at what is now the edge of the continental shelf or platform about 100 miles east of the present coast line. That this greater altitude was before the Ice age is proved by the fact that the inner Hudson channel now contains much glacial debris filling. That all of New York State was then higher than now is quite certain because, for example, with the lower Hudson region considerably elevated, the upstate region must also have been elevated (though possibly not so much) in order to maintain the gradients of the actively eroding streams.

*To summarize briefly the drainage and physiography of the State during the Tertiary, we may say that, with a certain few important exceptions, the major features as we see them today were practically the same toward the close of the Tertiary, and that these relief features were developed by erosion which began with the uplift of the great peneplain at the opening of this period or the close of the one just preceding. A few of the more notable differences between the drainage of the late Tertiary and the present are as follows: Very few, if any, lakes, waterfalls, or gorges existed; Lakes Erie and Ontario were absent and these basins contained important streams which appear to have drained westward into the Mississippi; the St Lawrence river probably had its source in the Thousand Islands region; the Mohawk river had its source on the divide at Little Falls, while the so-called Rome river flowed westward from Little Falls; West Canada creek entered the Rome river; the Sacandaga river entered the Mohawk; the State, especially the southeastern portion, was notably higher (perhaps not less than 2000 feet) than it now is so that the Atlantic coast line was about 100 miles farther out where the Hudson emptied into the ocean; and Long and Staten islands did not then exist as such.*

---

[1] It has been suggested by Chamberlain and Salisbury (Geology, vol. 1, page 529) that the very end of the Hudson, and other submerged channels, may have been deepened by tidal scouring and, if so, the figure (2000 feet) generally given may be too high. At any rate the Hudson channel at the Highlands is submerged nearly 800 feet which certainly implies an altitude of more than 1000 feet greater than now when the river was actively eroding.

## QUATERNARY PERIOD, INCLUDING NEW YORK IN THE GREAT ICE AGE

**The fact of the Ice age.** The Quaternary is the last great period of earth history, and it still continues for it has led up to the present-day conditions. This period was ushered in by the spreading of vast ice sheets over much of northern North America and Europe, which must take rank as one of the most interesting and remarkable occurrences of geological time. On first thought the existence of such vast ice sheets seems unbelievable, but the Ice age occurred so short a time ago that the records of the event are perfectly clear and conclusive. The fact of this great Ice age was discovered by Louis Agassiz in 1837, and fully announced before the British Scientific Association in 1840. For some years the idea was opposed, especially by advocates of the so-called iceberg theory. Now, however, no important event of earth history is more firmly established and no student of the subject ever questions the fact of the Quaternary Ice age.

Some of the proofs of the former presence of the great ice sheet are as follows: (1) polished and striated rock surfaces (see plate 39) which are precisely like those produced by existing glaciers, and which could not possibly have been produced by any other agency; (2) glacial boulders or " erratics " which are often somewhat rounded and scratched, and which have often been transported many miles from their parent rock ledges; (3) true glacial moraines, especially terminal moraines, like that which extends the full length of Long Island and marks the southernmost limit of the great ice sheet; and (4) the generally widespread distribution over most of the glaciated area of heterogeneous glacial debris, both unstratified and stratified, which is clearly transported material and typically rests upon the bedrock by sharp contact.

**Ice extent and centers of accumulation.** The best known existing great ice sheets are those of Greenland and Anarctica, especially the former which covers about 500,000 square miles. This glacier is so large and deep that only an occasional high rocky mountain projects above its surface, and the ice is known to be slowly moving outward in all directions from the interior to the margins of Greenland. Along the margins, where melting is more rapid, some land is exposed, but often the ice flows out into the ocean where it breaks off to form large icebergs.

The accompanying map (figure 29) shows the area of nearly 4,000,000 square miles of North America covered by ice at the

time of maximum glaciation, and also the three great centers of accumulation and dispersal of the ice. The directions of flow of the ice from these centers have been determined by the study of the directions of a very large number of glacial striae or scratches.

FIG. 29 Map of North America showing the maximum extent of the great ice sheets of the Quaternary period. The three great ice centers are shown by the letters as follows: L = Laborador or Laurentide Glacier; K = Keewatin Glacier; and C = Cordilleran Glacier. All of New York State, except probably the very southern border of Long Island and the southern part of Cattaraugus county, was buried under the ice.

It was the Labradorean or Laurentide ice sheet which spread southward over New York to cover all the State except the southern border of Long Island. It must of course be remembered that the

north Atlantic coast line was then considerably farther out than now because of the greater elevation of the land.

**Direction of movement and depth of ice in New York.** The fact that glacial ice flows as though it were a viscous substance is well known from studies of present-day glaciers in the Alps, Alaska, or the Greenland ice sheet. A common assumption, either that the land at the center of accumulation must have been thousands of feet higher or that the ice there must have been immensely thick, in order to permit flowage so far out from the center, is not necessary. For instance, if one proceeds to pour viscous tar slowly in one place upon a perfectly smooth (level) surface, the substance will gradually flow out in all directions, and at no time will the tar at the center of accumulation be very much thicker than at other places. The movement of the ice from one of the great centers was much like this, only in the case of the glacier the accumulation of snow and ice was by no means confined to the immediate centers of accumulation.

When the Labradorean ice sheet spread out southward as far as northern New York, the Adirondack mountains stood out as a considerable obstacle in the path of the moving ice, and the tendency was for the current to divide into two portions, one of which passed southwestward up the low, broad St Lawrence valley, and the other due southward through the deep, narrow Champlain valley. As the ice kept crowding from the rear, part of the St Lawrence ice lobe pushed into the Ontario basin, while another portion pushed its way up the broad, low Black river valley and finally into the Mohawk valley. At the same time the Champlain ice lobe found its way into the upper Hudson valley, and sent a branch lobe up the broad, low Mohawk valley. The two Mohawk lobes, the one from the west and the other from the east, met in the Mohawk valley not far from Little Falls. As the ice sheet continued to push southward, all the lowlands of northern New York were filled, a tongue or lobe was sent down the Hudson valley, and finally the whole State, except the southern border of Long Island, was buried under the ice. The general direction of ice movement at this time of greatest ice extent was southward to southwestward with perhaps some undercurrents determined by the larger topographic features. Thus we learn that the major relief features of the State very largely determined the direction of ice currents, except at the time of maximum glaciation when only the undercurrents were controlled.

These ideas are abundantly borne out by the character and distribution of the glacial striae and boulders over the State. Central

New York is literally strewn with thousands of glacial boulders or erratics which were transported from the Adirondacks by the ice, and similar boulders are occasionally found as far south as Binghamton or well down the Hudson valley. Evidences of glaciation also occur high up in the Adirondacks and the Catskills, so that the greatest depth of ice over New York State could not have been less than several thousand feet. In fact, we have every reason to believe that the Adirondacks, if not the Catskills, were completely buried. The reader may wonder how an ice sheet a mile thick in northern New York could have thinned out to disappearance at or near the southern border of the State, but observations on existing glaciers show that it is quite the habit of extensive ice bodies to thin out very rapidly near the margins, thus producing steep slopes along the ice fronts.

**Successive ice invasions.** The front of the great ice sheet, like that of ordinary valley glaciers, must have shown many advances and retreats. In the northern Mississippi valley, however, we have positive proof for several (perhaps five or six) important advances and retreats of the ice which gave rise to true interglacial stages. The strongest evidence is the presence of successive layers of glacial debris, a given layer often having been oxidized, eroded, and covered with vegetation before the next (overlying) layer was deposited. In drilling wells through the glacial deposits of Iowa, for example, two distinct deposits or layers of vegetation are often encountered at depths of from 100 to 200 feet. Near Toronto, Canada, plants which actually belong much farther south in a warmer climate have been found between two layers of glacial debris. Thus we know that some, at least, of the ice retreats produced interglacial stages with warmer climate and were sufficient greatly to reduce the size of the continental ice sheet or possibly to cause its entire disappearance.

In New York State no very positive evidence has as yet been found to prove truly multiple glaciation, though some phenomena as, for example, certain buried gorges, are very difficult to account for except on the basis of more than one advance and retreat of the ice. At any rate, there appears to be no good reason whatever to believe that there were more than two advances and retreats of the ice over the State, and for our purpose in considering only the general effects of glaciation we may practically disregard the problem of multiple glaciation because the final effects would have been essentially the same as a result of a single great glacial advance and retreat.

**Ice erosion.** Ice, like flowing water, has very little erosive effect upon rocks unless it is properly supplied with tools. When flowing ice is shod with hard rock fragments the power to erode is often pronounced because the work of abrasion is mostly accomplished by the rock fragments rather than by the soft ice itself. For instance, when the great ice lobe moved up the St Lawrence valley it was shod with many pieces of hard Precambric rocks, and the effects of erosion are remarkably well shown in the Thousand Islands region. Thus, about two miles due south of Clayton the writer has seen a succession of great grooves, covering an area of several acres, and cut into the hard, fresh Potsdam sandstone on top of a low hill. A little search will reveal polished and scratched or grooved rock surfaces in almost any part of the State. Granite ledges in the Adirondacks are often glaciated, and the freshness and hardness of the surface rock proves that the ice eroded all the deep preglacial soil as well as the zone of rotten rock, and an unknown amount of live or fresh rock.

In former years a very great erosive power was ascribed to flowing ice, but today some glacialists consider ice erosion to be almost negligible, while many others maintain that, under favorable conditions, flowing ice has a very considerable erosive effect. During the very long preglacial time, rock decomposition must have progressed so far that rotten rock, including soils, had accumulated to considerable depths, as today in the southern states. Such soils are called "residual" because they are derived by the decomposition of the very rocks on which they rest. But now one rarely ever sees rotten rock or soil in its original place in New York because such materials were nearly all scoured off by the passage of the great ice sheet, mixed with other soils and ground up rock fragments and deposited elsewhere. Such are called transported soils. Along the southern side of the State, where the erosive power of the ice was least, rotten rock is not so uncommonly seen.

Ice, shod with hard rock fragments and flowing through a deep, comparatively narrow valley of soft rock, is especially powerful as an erosive agent because the abrasive tools are supplied; the work to be done is easy; and the increased depth of the ice where crowded into a deep, narrow valley causes greater pressure on the bottom and sides of the valley. Many of the valleys of northern New York were thus very favorably situated for ice erosion, as for example, the Champlain, St Lawrence, Black river, and Finger lakes valleys, as well as many of the nearly north-south valleys of the Adirondacks. The writer has made a special study of ice

erosion in the Black river valley, and figure 35 is a structure section across the valley showing the rock terraces and the relations of the various rock formations. The high, steep, terrace fronts are certainly young topographic features which could not have been present at the close of the long preglacial erosion period, nor could they have been formed since the Ice age because glacial deposits, even near the valley bottom, have not yet been removed. There is still the possibility that glacial waters may have done the work, but there is no evidence for such vigorous water action especially on the higher part of the Trenton limestone terrace where records would surely be left. On the contrary, there are glaciated rock surfaces and also glacial deposits (kames) on the great limestone terrace and near the base of the steep front of the shale terrace, so that the work could not have been done by glacial waters before the ice retreat. Evidently, we have here a fine example of ice erosion, and before the Ice age the limestones and shales extended somewhat farther eastward than they now do. The conditions for ice erosion were here unusually favorable because the ice, in its great sweep around the Adirondacks, was shod with many fragments of very hard rocks and entered the deep Black river valley striking with greatest force against the soft sedimentary rocks of the west side of the valley. As the figure clearly shows, the very soft shales were worn back more than the harder limestones, while the very hard Precambric rocks were very little affected. This is perhaps the best example of ice erosion in northern New York, and even here we must admit that only soft rocks were much eroded and that the great preglacial Black river valley was comparatively little modified. If soft shales had made up the valley bottom, ice erosion would have caused considerable deepening as was, no doubt, the case in the valleys of the Finger lakes region.

Most of the Adirondack mountain peaks, especially the more isolated ones, were thoroughly scraped off and rounded down to the very live or fresh rock (see upper view, plate 17), while the favorably situated valleys were vigorously glaciated by the removal of all the rotten and at least some of the fresh rock, especially when this latter was the comparatively soft Grenville limestone. Such phenomena are particularly well exhibited in Warren county (see figure 13) where the landscape is characterized by many great, glaciated rock domes which rise above the valleys of weak Grenville. In a few cases where the ice moved directly across deep valleys, like that between Lake George village and Warrensburg, the rotten rock to great depth may still be seen in its original place.

In conclusion we may say that while many comparatively small, local features were produced by ice erosion, the major topographic features of the State were practically unaffected by ice erosion due to the passage of the great ice sheet.

**Ice deposits.** The vast amount of debris transported by the great ice sheet was carried either on its surface, or frozen within it, or pushed along under it. It was very heterogeneous material ranging from the finest clay through sand and gravel, to boulders of many tons weight. The deposition of these materials, as we now see them, took place during both the advance and retreat of the ice, but chiefly during its retreat. Most of the deposits made during the ice advance were obliterated by ice erosion, while those formed during the ice retreat have been left intact except for the small amount of postglacial erosion. The general term applied to all deposits of glacial origin is "drift," this term having been given at the time when they were regarded as flood deposits. Drift deposits cover practically all of New York State except where bare rock is actually exposed, and its thickness is very variable, ranging from nothing to several hundred feet.

The ice sheet could advance only when the rate of motion was greater than the rate of melting of the ice front and vice versa in the case of retreat. Thus it is true, though seemingly paradoxical, to assert that the ice was constantly flowing southward even while the ice front was retreating northward. Whenever, during the great general retreat, the ice front remained stationary because the forward motion of the ice was just counterbalanced by the melting, all the ice reaching the margin of the glacier dropped its load to build up a *terminal moraine.* Such a moraine is a more or less distinct range of low hills and depressions consisting of very heterogeneous and generally unstratified debris, though at times waters emerging from the ice caused stratification. The depressions are usually called *kettle holes.* The so-called great terminal moraine marks the southernmost limit of the ice sheet, and is wonderfully well shown by the ridge of low irregular hills extending the whole length of Long Island (see plate 12). It is also clearly traceable across northern New Jersey and Pennsylvania and passes through southern Cattaraugus county in New York. Terminal moraines farther northward are generally not so long nor sharply defined, the one of perhaps most prominence having been traced from Herkimer through Oriskany Falls, Cortland, Watkins, Bath, Portageville, Dayton, and Jamestown. Moraines, either terminal or lateral, are often locally very prominently developed.

When the ice front paused for a considerable time upon a rather flat surface, the debris-laden streams emerging from the ice formed what is called an *overwash plain* by depositing layers of sediment over the flat surface. The finest illustration of such an overwash plain in the State is all of that part of Long island lying just south of the great terminal moraine, and known as the Jamaica plain toward the east (see plate 12).

When the ice front extended across a more rugged country, with valleys sloping away from the ice, the large glacial streams, heavy laden with debris, caused more or less deposition of materials on the valley bottoms often for many miles beyond the ice front. Such deposits, known as *valley trains,* are especially well developed along most of the large south-flowing tributaries of the upper Susquehanna river in the Southwestern plateau province.

*Glacial boulders, or erratics,* have already been referred to; they are simply blocks of rock or boulders from the top of the ice or within it which have been left strewn over the country as a result of the melting of the ice. They vary in size from small pebbles to those of many tons weight (see plate 40), and are naturally most commonly derived from the harder and more resistant rock formations. Thus erratics from the Adirondacks are very numerous in east-central New York, some having even been transported to the southern border of the State. Erratics are often found high up on the mountains, and sometimes they have been left stranded in remarkably balanced positions.

A very extensive glacial deposit, called the *ground moraine,* is simply the heterogeneous, typically unstratified, debris from the bottom of the ice which was deposited, sometimes during the ice advance, but most often during its melting and retreat. When it is mostly very fine material with pebbles or boulders scattered through its mass, it is known as *till* or *boulder clay*. The pebbles or boulders of the till are commonly faceted and striated as a result of having been rubbed against underlying rock formations.

Another type of glacial deposit of unusual interest is the *drumlin* which is, in reality, only a special form of ground moraine material or till. The typical drumlins of New York State are low, rounded mounds of till with elliptical bases and steeper slopes on the north sides and with long axes parallel to the direction of ice movement (see plate 42). In height they rarely exceed 200 feet, being most often less than 100 feet. The origin of the drumlins has not yet been satisfactorily determined, though it is known that they formed near the margin of the ice either by the erosion of an earlier drift

layer, or by accumulation beneath the ice under peculiarly favorable conditions, as perhaps along longitudinal crevasses or fissures. One of the finest and most extensive exhibitions of drumlins in the world is the region of western New York from Oswego and Syracuse to west of Rochester. Thousands of drumlins there rise above the general level of the Ontario plain, the New York Central Railroad, from Syracuse to Rochester, passing through the very midst of them.

Another type of glacial deposit in the low hill or hillock form is the *kame* which, in contrast with the drumlin, always consists of stratified drift. Kames are seldom as much as 200 feet high, and typically they have rounded bases though frequently they are very irregular in shape. At times they exist as isolated masses or hills or in small groups, while often they are associated with the unstratified deposits of the moraines. When grouped, deep depressions occur between the hills to form what is called the knob and kettle structure. Kames were formed at or near the margin of the retreating ice, and so are found in all parts of the State. They most generally occur in valley bottoms, but sometimes on hillsides or even hilltops. They are especially common along the line of the great terminal moraine (for example, on Long island), and also along the line of the important terminal moraine already described from central to western New York. For example, in the vicinity of Oriskany Falls kames are so numerous as to form a striking feature of the landscape in the Oriskany valley. They were formed as deposits by debris-laden streams emerging from the margin of the ice, the water sometimes having risen like great fountains because of pressure. Such deposits are now actually in process of formation along the edge of the great Malaspina glacier of Alaska.

During the retreat of the ice, glacial lakes were numerous, especially after the ice front had passed north of the Susquehanna-Allegany divide because the north-sloping valleys were dammed by the ice thus ponding the waters in the valleys. Some materials were directly deposited from the glacier in those lakes, but more was brought in by debris-laden streams flowing from the land already freed from the ice. Such glacial lakes and their deposits are common and of unusual interest, but they will be described under a subsequent heading.

In conclusion we may say that the deposition of glacial materials, like glacial erosion, has not changed the major topographic features of the State. The general tendency of ice deposits has been to fill or partially fill depressions and thus to diminish the ruggedness of the topography.

**Great Lakes history.** The Great Lakes certainly did not exist before the Ice age, but instead the depressions in that region were occupied by stream channels. During the very long erosion period (already discussed) from the Paleozoic to the Cenozoic, no lakes, except possibly a few very small ones due to landslides, beaver dams, etc., could have existed. Compared with such an immense length of time lakes are, at most, only ephemeral features of the earth's surface because they are soon destroyed either by being filled with sediments, or by having their outlets cut down, or both. Since the Great Lakes are of postglacial origin it is, then, proper to ask how they came into existence. During preglacial time, as we have learned, broad valleys were cut out along belts of weak rock in the Great Lakes region, and these old valleys, to a considerable extent at least, account for the present depressions, but not for the closed lake basins. This idea of preglacial stream valleys is not at all opposed by the fact that some of the lake bottoms are now well below sea level because there has been a notable subsidence of the region since preglacial time. The surface of Lake Erie is 573 feet and its deepest point 369 feet above sea level, while the surface of Lake Ontario is 247 feet above and its deepest point is 491 feet below sea level. The greatest depth (738 feet) of Lake Ontario is well toward the east end and not far from the south shore, and if we consider this deep place as due to preglacial erosion, we ought to find an outlet channel. But no such outlet channel exists because the whole eastern end, at least, of the lake is certainly rock-rimmed. As Tarr has said: "There could hardly be a valley over 700 feet deep and broad enough to form the continuation of the preglacial Ontario valley, which is so completely obscured by drift that not the least trace of it has been found on the surface."[1] To assume that this deep part of the basin was produced by warping of the land is not borne out by examining the exposed strata on all sides. It therefore seems quite certain that the preglacial Ontario depression was here considerably deepened by ice erosion. The conditions were very favorable for such erosion because the rocks were chiefly soft Ordovicic shales; because the ice flowed through a deep preglacial valley; and because there was unusual crowding of ice into this valley due to the pronounced deflection of a great ice current around the Adirondacks on the west side. Strong arguments might be adduced to show that by ice erosion, portions, at least, of all the

[1] Tarr's Physical Geography of New York State, p. 235.

lake basins were appreciably deepened. Even so, however, we have not yet accounted for the present closed basins. In the writer's opinion the two most important phenomena which have contributed to the formation of the closed basins of the Great Lakes are the great drift accumulations along the south side and the tilting of the land downward on the north side of this region. The deep drift deposits must certainly have been very effective in damming up the south or southwesterly-flowing preglacial streams of the region. For example, the deep channel of the so-called Dundas

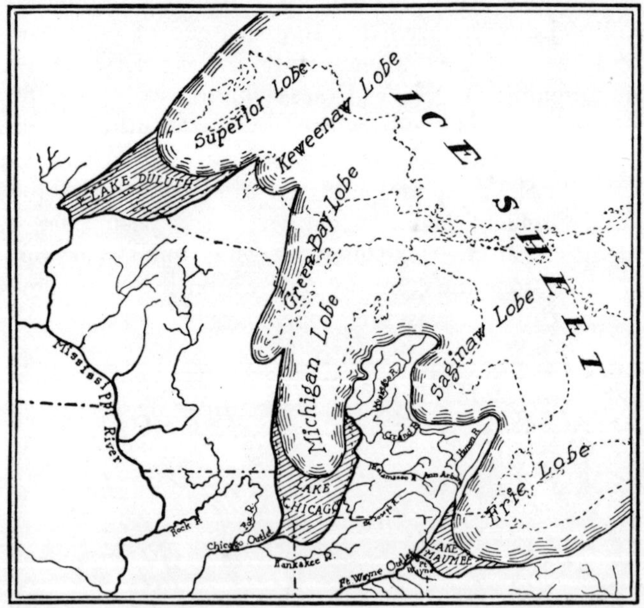


FIG. 30 The first stage in the formation of the Great Lakes, when most of the region was still buried under the ice.
After Taylor & Leverett

river (see figure 27) has been drift-filled as proved by many well borings, and a distinct moraine extends around the southern half of Lake Michigan. The great dumping ground of ice-transported materials from the north was in general along the southern side of the Great Lakes and southward. Late in the Ice age the land on the northern side of the Great Lakes region was lower than it is today as proved by the tilted character of certain well-known beaches of extinct glacial lakes (see below). Such a differential tilting or warping of the land must have helped to form the closed basins by tending to stop the southward or southwestward drainage

from the region. To summarize, we may say that the present Great Lakes basins are due to a combination of factors, the more important of which were: the formation of preglacial valleys by stream erosion; a more or less deepening of these valleys by ice erosion; the great accumulation of glacial debris along the southern side of the Great Lakes region; and the tilting of the land downward toward the north.

We are now ready to trace out the principal stages in the history of the Great Lakes during the final retreat of the ice sheet. When the ice front had receded far enough northward to uncover the western end of Lake Superior, the southern end of Lake Michigan, and an area west of the present end of Lake Erie, small lakes were formed against the ice walls (see figure 30). One of these has been called Lake Duluth which drained southward into the Mississippi; the second Lake Chicago which drained past Chicago through the Illinois river and into the Mississippi; and the third Lake Maumee which drained southwestward past Fort Wayne through the Wabash river and into the Ohio and Mississippi.

At a still later stage the conditions shown on the map (figure 31) existed. Lake Chicago was then much larger, and Lake Maumee

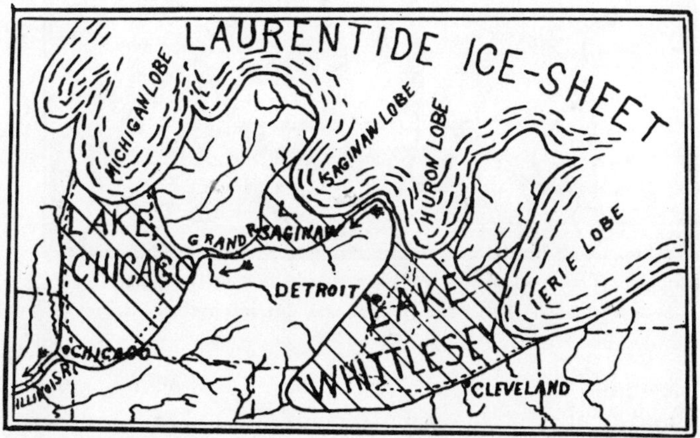


FIG. 31 A later stage of Great Lakes history, showing how the eastern and western ice margin lakes combined with outlet past Chicago.

After Taylor

had expanded into the extensive Lake Whittlesey which covered nearly all of Lake Erie as well as the immediately surrounding country. Lake Whittlesey was at a lower level than the former Maumee and the outlet past Fort Wayne ceased, but the drainage

from Whittlesey was westward by a large river flowing through small Lake Saginaw and into Lake Chicago, which latter still emptied through the Illinois river.

At a still later stage (figure 32) Lake Saginaw merged with the waters of the Erie basin to form the large Lake Warren which extended along the ice front eastward nearly to central New York. As the map clearly shows, the Finger lakes basins of New York were then occupied by Warren waters, while Niagara Falls were not then in existence because that region was also covered by Lake Warren. Lake Warren continued to discharge westward into Lake Chicago and the Mississippi river until a very late stage, when the waters had worked their way along the border of the Ontario ice

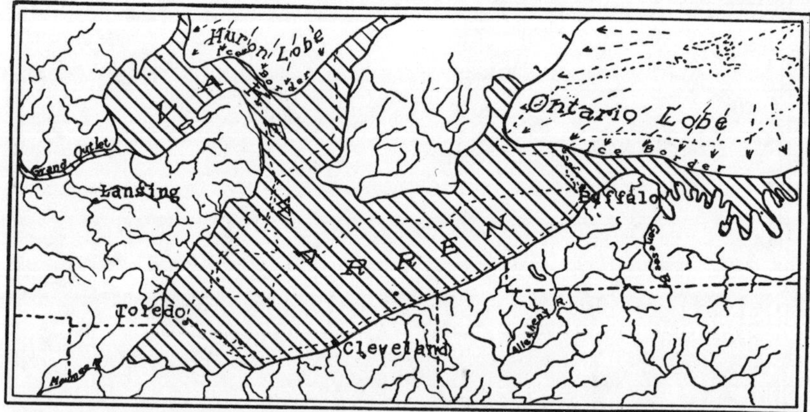


FIG. 32 Glacial Lake Warren. At this stage the discharge of the lake was still westward to Lake Chicago and the Mississippi river, while the eastern end of the lake covered most of the Finger Lakes region of New York.

Modified from Taylor & Leverett, U. S. G. S.

lobe into the Mohawk valley which was then occupied by a large glacial lake (held up by the Ontario ice lobe on the west and the Champlain-Hudson lobe on the east) and thence into the Hudson valley. Thus, for the first time, the Great Lakes drainage passed eastward into the Atlantic ocean. This great volume of water draining eastward was often in the form of distinct streams with the ice front for north wall and the high land of the Helderberg escarpment for wall on the south. Many of these glacial stream channels, which are still plainly visible, have been studied and mapped by Professor Fairchild.

By successive stages, due to a complete removal of ice from central New York and a draining of the glacial lake in the Mohawk valley, the waters dropped to below Warren level until Lake Iroquois was formed (see figure 33). The old beach line of this lake is still plainly visible in New York and with some slight interruptions has been traced from near the mouth of Niagara river to just north of Rochester, past Syracuse, along the south, east, and north sides of Oneida lake, and thence along the western base of the Tug Hill plateau to near Watertown. The well-known ridge road between Niagara river and Rochester is built on the old

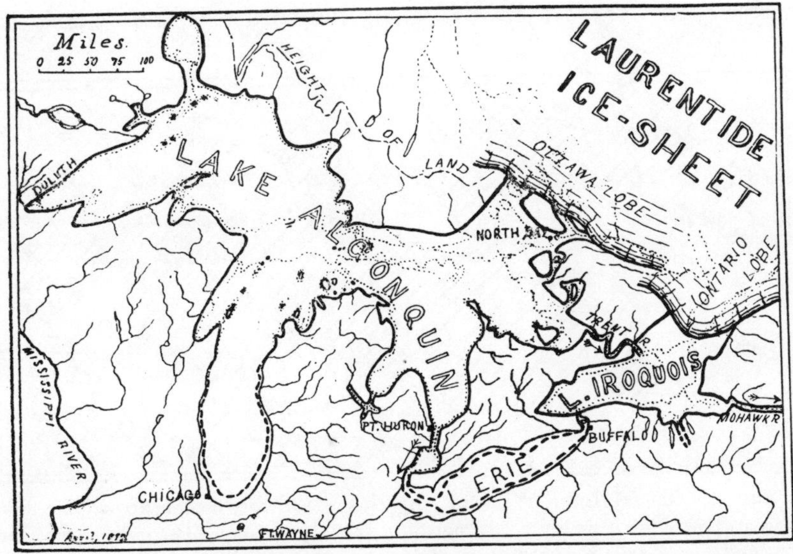


FIG. 33 The Algonquin-Iroquois stage of Great Lakes history when the ice had retreated far enough to open the outlet through the Mohawk valley.

After Taylor

Iroquois beach deposit. Lake Iroquois covered somewhat more than the present area of Lake Ontario, and the distinctly lower water level here than in the Erie basin allowed the modern Niagara river to begin its history by flowing northward over the limestone plain near Buffalo. Meantime the waters of the upper lake basins had merged to form Lake Algonquin which at first probably discharged past Detroit through the Erie basin and into Lake Iroquois by way of Niagara river. Later, however, when the ice had withdrawn a little farther northward, a lower outlet was opened through the Trent river by which Lake Algonquin drained into Lake

Iroquois. We know that the old Trent river channel is now higher than the Detroit outlet, but some of the proofs for the existence of the Trent outlet are as follows: the presence there of a large, distinct river channel; the convergence of the beaches toward that channel; and the fact that the land was then considerably lower on the north or northeast side of Lakes Ontario and Erie than on the south side. For example, in following the old Iroquois beach we find that it now gradually rises to higher levels until, even at Watertown, it is several hundred feet higher than near the mouth of the Niagara river. This tilting of the beach has been due to raising of the land since the lake existed, and it is evident therefore that during the Algonquin-Iroquois stage the Trent river channel was lower than that past Detroit. During this Lake Iroquois stage the waters of all the Great Lakes region discharged through the Mohawk-Hudson valleys, and the volume of water which flowed past Rome, Utica, and across the preglacial divide at Little Falls must have been as great, if not greater, than that which now goes over Niagara Falls. Much of the gorge cutting at Little Falls was accomplished by this great volume of water. The St Lawrence valley was still buried under the ice.

Still later the ice withdrew enough to allow the Algonquin-Iroquois waters to discharge along the northern base of the Adirondacks and into what appears to have been ice-ponded waters in the Champlain basin, and thence southward into the Hudson valley. The Mohawk river outlet was thus abandoned.

Finally the ice retreated far enough to free the St Lawrence valley when the waters of the Great Lakes region dropped to a still lower level, bringing about the Nipissing Great Lakes stage (see figure 34). The Nipissing lakes found a low outlet through the Ottawa river (then free from ice) and into the Champlain arm of the sea. Postglacial warping of the land brought the Great Lakes region into the present condition, but this, and the Champlain subsidence, being really postglacial features will be described toward the end of the chapter.

**Other existing lakes and their origin.** Counting all, from the smallest to the largest, there are within the borders of New York State thousands of lakes, which constitute one of the most striking differences between the geography of the present and that of preglacial time. These lakes are widely scattered over the State though there are three general regions worthy of particular mention as follows: the Finger lakes region of western New York; the Adirondack mountains; and the southeastern portion of the State.

The linear type of lake is by far the most common, this being preeminently true of the Finger lakes and to a large extent of the Adirondacks. It is well known that most of the larger lakes, especially those of the linear type, occupy portions of preglacial stream channels. All the existing lakes are due, either directly or indirectly, to glacial action, and among the ways by which such bodies of water were formed are these: by building dams of glacial drift across old river channels; by ice erosion; and by the filling of the numerous depressions which were formed by irregular deposition of the drift (kettle holes, etc.). Hundreds of small lakes, often

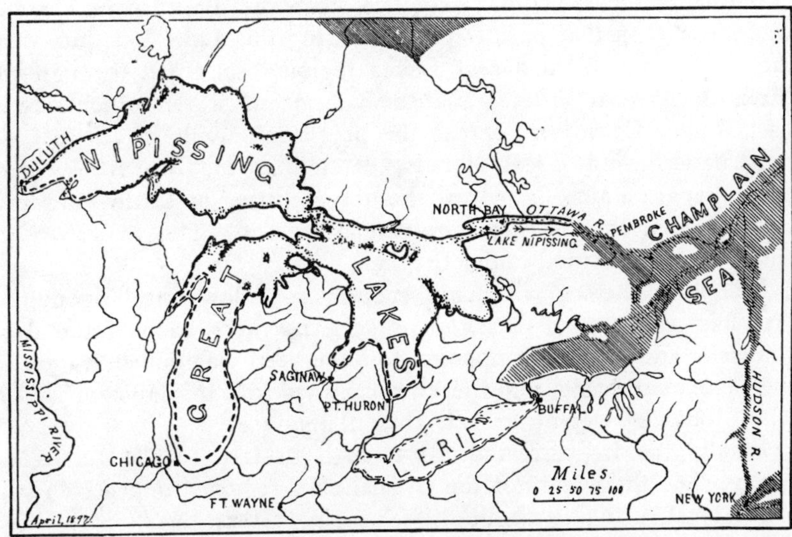


FIG. 34 The time of the Nipissing Great Lakes and Champlain submergence. The shaded area on the east was covered by sea water.

After Taylor

not more than mere ponds in size, belong to the last named type, while most of the large lakes are due chiefly to the existence of drift dams.

Much has been written concerning the origin of the Finger lakes, and only the briefest summary will here be given. All are agreed that the lakes of this remarkable group occupy preglacial valleys, most of which, at least, contained north-flowing streams. These lakes have dams of glacial drift across their lower (north) ends, and the dams have largely contributed to the formation of the lakes, being in some cases perhaps the sole cause of the lakes. In the

cases of the two largest lakes, Seneca and Cayuga, there is, however, strong evidence that the preglacial channels were notably deepened by ice erosion.[1] As Professor Tarr says: "They offered broad channel ways, along which the ice streams moved much more easily than upon the neighboring irregular hilltops. Not only was the movement more rapid, but the depth of ice was greater. The position of the rocks, dipping southward, and the nature of the friable shales conspired toward rapid erosion; and so these north and south preglacial valleys were markedly deepened. Evidence of this comes from the side streams. The rock bottoms of the preglacial valleys of these tributary streams are not now below the level of the lake water in the southern part of the valley (Cayuga). If all the drift could be removed and the streams be allowed to flow along the line of the course of the preglacial valleys and enter the valley of Lake Cayuga as it now stands, excepting that it be robbed of water, they would tumble between 300 and 400 feet in a distance of about a mile, commencing their descent near the present lake margin, a most unnatural condition for mature tributaries near their mouth."[2] Thus it appears quite certain that the preglacial Cayuga and Seneca valleys, at least, were notably deepened by ice erosion below the level of the mouths of the preglacial tributary streams.

Most of the numerous Adirondack lakes have certainly been formed by irregular damming of preglacial valleys by glacial drift. It is quite the rule to find the outlets of these lakes flowing through such loose materials. By ice erosion many of the favorably situated valleys were no doubt somewhat modified, but up to the present time we have no good example of a lake basin produced by that agency. The hard Precambric rocks were not so easily eroded by the ice. Attention is called to the prominent lake belt in the middle of the Adirondack province, and running in a north-northeast by south-southwest direction. This belt comprises many well-known lakes as Placid, Saranac, Tupper, Long, Blue Mountain, Big Moose, and Fulton Chain lakes. Sometimes small lakes or ponds are situated well toward mountain tops because of favorably located drift deposits. A good example of such lake lies at an altitude of 2620 feet, and well toward the top of Crane mountain in Warren county.

---

[1] The surfaces of Seneca and Cayuga lakes are respectively 444 and 381 feet above sea level, while their deepest places are respectively 186 and 119 feet below sea level.

[2] Tarr's Physical Geography of New York State, p. 181–82.

Many of the existing Adirondack lakes were formerly of larger extent as proved by delta deposits above the present lake levels. Two lakes of this class recently coming under the writer's observation are Schroon lake in Warren-Essex counties, and Piseco lake in Hamilton county. The water of Schroon lake was once fully 70 feet higher when it extended some eight or ten miles farther up the Schroon river, with a branch reaching over the area of the present Paradox lake, and also for some six or eight miles farther southward to cover all the lowland around Chestertown, and with a prominent branch extending over the area of the present Brant lake. Piseco lake was at one time clearly twenty feet higher, and then extended several miles farther northward.

The valley of Lake Champlain was favorably situated for ice erosion, and it bears evidence of having been vigorously glaciated though it has not been proved that the existing closed basin is chiefly due to ice erosion. At the close of the Ice age, tide water entered the valley. The present lake basin is due principally to a combination of late elevation of the land, with greater uplift on the north; heavy glacial accumulations toward the north; and possibly some deepening as a result of ice erosion.

Lake George is justly famous because, from the standpoint of length and depth in proportion to width, no other lake in the State occupies such a remarkable depression. This depression has been determined by ordinary erosion along lines of prominent faults. There was a preglacial divide where the "Narrows" are now located, and this divide appears to have been considerably lowered by ice erosion when part of the Champlain ice lobe plowed its way through the deep, narrow valley. The waters are now held in by glacial deposits at each end.

In southeastern New York, from the Connecticut state line westward to the southern Catskills in Sullivan county, there are many lakes, though all are comparatively small. With few exceptions these lakes appear to be of the usual drift dam type. Greenwood lake, at an altitude of 621 feet and passing from Orange county across the state line into New Jersey, is the largest in this part of the State. Three small lakes near the summit of Shawangunk mountain, and close to its eastern edge, deserving special mention are: Mohonk, Minnewaska, and Awosting. Mohonk lake, which is so widely known both because of its remarkable situation and as a place where so many peace conferences have been held, may be taken as the type of the three. The altitude of this lake is more than 1200 feet or about 1000 feet above the base of the mountain

ridge on which it is located. It is almost completely surrounded by walls of hard Shawangunk conglomerate, while the lake basin itself is in the soft underlying Ordovicic shales. This lake does not appear to owe its origin to a dam of glacial drift, but rather to ice erosion in the soft shales at a place where they had already been exposed to view before the oncoming of the ice. Such patches of shale occur at several places on the mountain.

Near the very western end of the State lies another lake remarkably situated. This is Lake Chautauqua, famous as the great center of Chautauqua assemblies. The altitude of the lake is 1338 feet, and its northern end is near the edge of the steep front of the Southwestern plateau province where it overlooks the low Erie plain. The drainage is southward into the Allegheny river, but the narrow place near the middle of the lake strongly suggests a preglacial divide there. As Tarr says: "If this view be true, Chautauqua lake is made up of parts of two valleys, one north-sloping, the other south-sloping, and each dammed by heavy morainic accumulations."[1]

**Extinct glacial lakes.** Hundreds of extinct glacial lakes are known to be scattered over the State. Some of these existed only during the time of the ice retreat, while others persisted for a greater or lesser length of time after the Ice age. Lakes Warren, Iroquois etc., already described, were fine examples of the first type. North-sloping valleys were particularly favorable for the development of glacial lakes during the retreat of the ice because the ice front always acted as a dam across such valleys, thus allowing the waters to become ponded. When the ice front stood across the northern ends of the Finger Lakes valleys, the waters in those valleys were ponded at much higher levels than they now are, and the ancient water levels are more or less clearly marked by the old beach lines.

Perhaps the finest example of a large, wholly extinct glacial lake is Black lake, which occupied a good portion of the Black river valley on the western side of the Adirondacks. This lake, small at first, was formed by ponding the waters in the upper Black river valley around Forestport, Oneida county, in front of the waning (northward retreating) ice lobe in the Black river valley. Its first discharge was probably southward past Remsen. Further retreat of the ice lobe permitted an enlargement of the lake to the region around Boonville, and the discharge was then southward along the channel of the present Lansing kill. The deep, narrow gorge a

[1] Tarr's Physical Geography of New York State, p. 205.

few miles south of Boonville, along this stream, was mostly cut by the fairly large stream which drained the glacial lake at this stage. The southward discharge through this channel appears to have been into the Lansing Kill lake where a delta deposit was formed a few miles north of Rome and now the site of the Delta reservoir. Lansing Kill lake in turn drained through the Mohawk Valley. Still further retreat of the ice lobe allowed Black lake to expand greatly until it reached from the region around Forestport to north of Lowville, when it had a width of from five to ten miles. When the ice withdrew enough to permit a discharge of water around the north base of Tug hill, and into Lake Iroquois, the level of the lake rapidly fell until the ice barrier was completely removed. The former presence of this great glacial lake is conclusively shown by the extensive development of unquestioned delta deposits. Where the streams from the Adirondacks, especially the larger ones such as Black, Moose, and Independent rivers, emptied into the lake, delta deposits were rapidly built up to near the lake surface because those streams were heavily charged with debris from the newly drift-strewn mountains. These delta deposits became more or less merged, and they show a remarkable concordance of altitudes over the sand flat or sand plain country on the east side of Black river. This great delta deposit is several miles wide; very flat-topped except where trenched by postglacial streams; presents a steep front toward the river; and shows a depth of from 200 to 250 feet along the western edge. Figure 35 clearly shows the relation of the delta deposit to the old rocks of the valley.

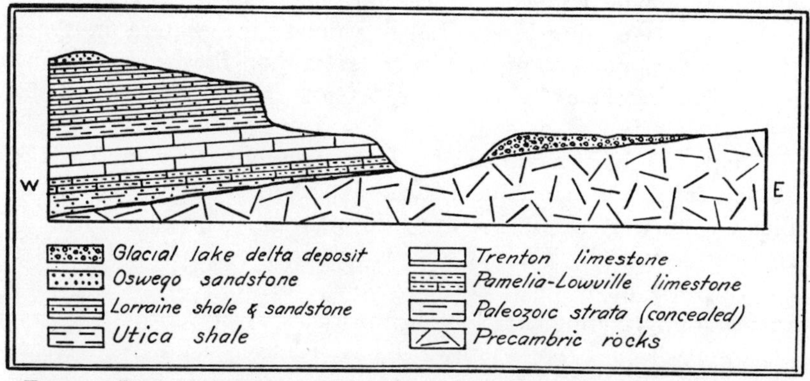


FIG. 35 East-west section across the Black river valley, 2½ miles north of Lyons Falls, showing the terraced character of the Paleozoic strata and their relations to the Precambric Adirondack rocks. On the east side, the position of the glacial lake delta deposit is shown. Length of section 12½ miles. Vertical scale greatly exaggerated.

After W. J. Miller, N. Y. State Mus. Bul. 135

In many cases where the edge of an ice lobe extended across the mouth of an east, or west, or even south-sloping valley, glacial lakes were formed. A fine example of a glacial lake (now extinct) formed in a south-sloping valley has been called glacial Lake Sacandaga which covered many square miles of the bottom of the broad, deep valley in which Johnstown, Gloversville, and Northville are located. Through this valley, which has a width of several miles and a maximum depth of over a thousand feet, the preglacial Sacandaga flowed southward into the Mohawk. During the general ice retreat, but when the Mohawk glacial lobe was still present, morainic deposits along the margin of the ice lobe formed an effective barrier across the mouth of the valley thus ponding the waters over the valley bottom and causing the Sacandaga to find an outlet northeastward over the low divide at Conklingville. The altitude of the lake corresponded approximately with the present 780 foot contour line, though it is quite certain that the land was then somewhat lower. This lake persisted for a good while after the disappearance of the ice because of the effective drift dam, and even today, in the spring of the year, a number of square miles of swamp in the lowest part of the valley are flooded. The lake was drained by cutting down the divide at Conklingville. It is interesting to note in passing that the construction of the proposed Sacandaga reservoir, by means of a dam at Conklingville, would almost exactly restore this former glacial lake.

Many other glacial lakes are known to have been formed by ponding of water alongside the waning Mohawk ice lobe. During the melting of the ice tongue from the Hudson and Champlain valleys, many small glacial lakes are also known to have been formed in the tributary valleys because of ice dams across them.

New York State fairly abounds in such extinct glacial lakes, and though comparatively few have yet been described, they are usually easily recognizable by means of the typical, flat-topped, delta deposits of crudely stratified sands, gravels and clays.

**Drainage changes, gorges, and waterfalls.** Along with its lakes, New York State is also famous for its numerous gorges and waterfalls, which are also largely due to the great Ice age. As a result of the very long preglacial erosion period, it is perfectly clear that typical, steep-sided, narrow gorges and true waterfalls must have been very uncommon, if present at all. Like lakes, such features are ephemeral because, under our conditions of climate, gorges soon (geologically) widen at the top and waterfalls disappear by retreat or by wearing away the hard rock over which they fall.

Drainage changes, aside from those already described in connection with the history of lakes, are also numerous in New York. It must be remembered that, with few exceptions (for example, the basins of Lakes Ontario and Erie, Niagara river, and possibly the St Lawrence river), the major drainage lines of the State were little changed during the Ice age because the principal valleys were mostly the same before and after glaciation. It is the present purpose briefly to describe only some of the most important and best known cases of stream changes due to the Ice age.[1]

From the standpoint of both geography and human history, the gorge at Little Falls is the most important in New York State (see figure 7 and plate 42 and also the description in chapter 2). Before the Ice age there was a divide, instead of the gorge, several hundred feet above the present river level, which consisted of hard Little Falls dolomite. The prominence of this rock barrier was greatly increased by the tilting of the strata due to the development of the Little Falls fault. The Mohawk river flowed eastward, and the now extinct Rome river flowed westward, from this divide (see figure 36). During the Ice age the divide was somewhat lowered

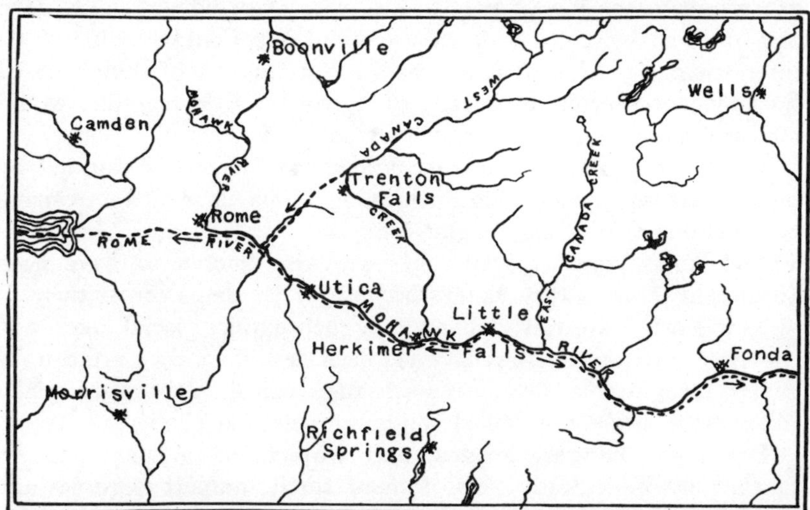


FIG. 36 Sketch map of central New York, showing the relation of preglacial to postglacial drainage. Preglacial streams shown by dotted lines only where essentially different from existing streams.

Based upon work of A. P. Brigham

[1] All the drainage changes now to be described will be much better understood by consulting the large government topographic maps of the regions considered.

by ice erosion, and during the Algonquin-Iroquois stage of the Great Lakes history, we have learned that these lakes discharged through the Mohawk valley and across the Little Falls divide. It was the passage of this great volume of water over the divide which caused the cutting of most of the gorge as we now find it, except for the narrow trench in the hard, underlying Precambric rock which is no doubt due to postglacial erosion. During the Iroquois stage an arm of the lake extended along the valley from Rome to Little Falls. All the streams from north and south which entered this arm of the lake were heavily charged with debris from the newly drift-covered regions and, the current not being strong enough to carry away the debris, the valley from Rome to Little Falls was built up (aggraded) to such an extent that, after the disappearance of Lake Iroquois, the drainage from Rome was able to continue eastward. Thus we have here a very fine example of exact reversal of drainage directly due to glaciation and by this means the upper waters of the Mohawk were added to the preglacial Mohawk.

Closely associated with the above is the postglacial history of West Canada creek and the famous chasm at Trenton Falls. The preglacial West Canada creek flowed from Prospect (upper end of Trenton chasm) past Holland Patent, through the valley of the present Nine Mile creek, and into the Rome river opposite the village of Oriskany. This channel was completely blocked by glacial drift at Prospect so that the creek was forced to find a new course southward over the limestone at Trenton Falls, and thence southeastward to its present mouth at Herkimer. The gorge, between Prospect and Trenton Falls villages, is 2½ miles long and from 100 to 200 feet deep, and has been cut into the Trenton limestone by the postglacial stream. It contains five or six waterfalls ranging in height from 10 to 126 feet, the total drop of the water in the 2½ miles of the gorge being 360 feet (see plates 43 and 44).

In the southeastern Adirondacks, the upper waters of the Hudson river present some very interesting examples of drainage changes. In fact, it is not too much to say that the larger drainage features of that region have been well nigh revolutionized as a result of glaciation. The accompanying sketch map (fig. 37) gives a fair idea of the changes, but reference to the State geologic map and to the topographic maps of the region is greatly to be desired. The State geologic map shows two distinct embayments of Paleozoic rocks forming valleys which extend northward, one to Northville and the other to Corinth, and into the mass of Precambric rocks of the

Adirondacks. It is certain that these valleys contained important preglacial streams which flowed southward out of the mountains. Now, however, the Sacandaga river enters the north end of the

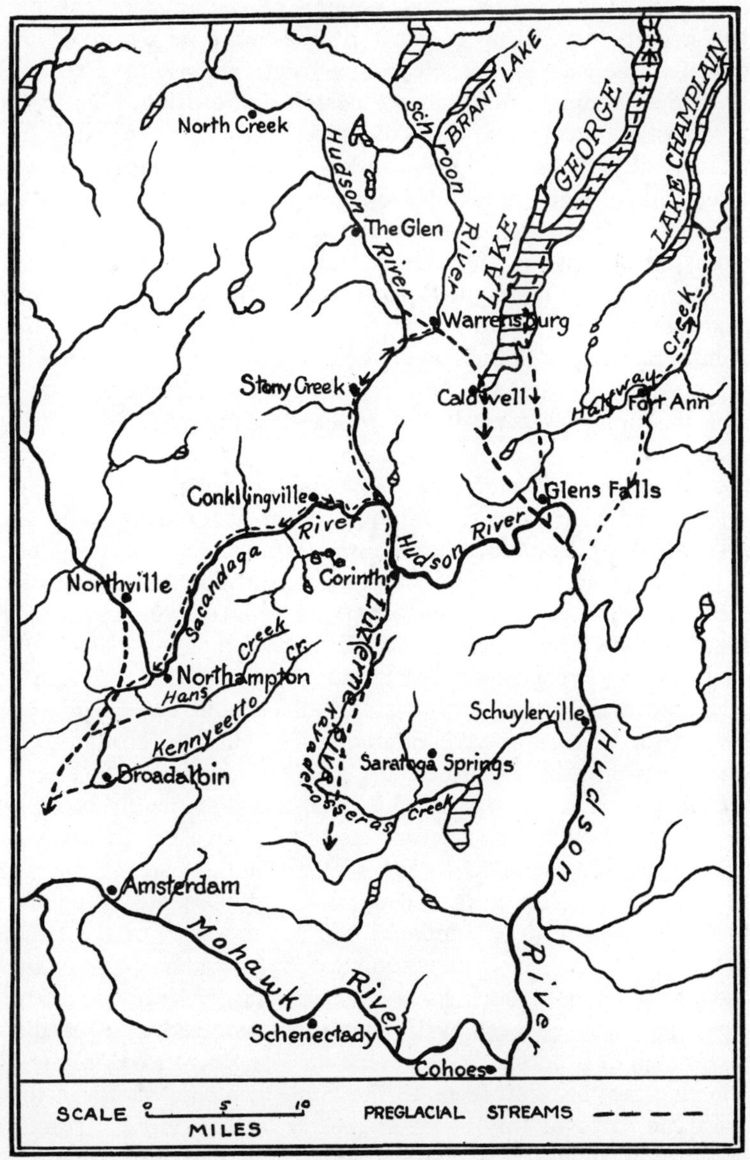


FIG. 37 Sketch map of the southeastern Adirondack region, showing the relation of the preglacial drainage to that of the present. Preglacial courses shown only where essentially different from present streams.

After W. J. Miller, Bul. Geol. Soc. Am., vol. 22

first named valley only to make a very sharp turn back on its course to flow across the mountains and into the Hudson at Luzerne. A preglacial divide was located at Conklingville as shown by the gorge there; the perfectly graded condition of the valley bottom westward from that place; and the flaring of the valley westward. This remarkable deflection of the river was caused by the building of a morainic blockade across the southern end of the Paleozoic rock valley from Broadalbin to Gloversville. The peculiar courses of Hans and Kennyetto creeks are thus also easily explained.

The Hudson river now flows through a gorge more than 1000 feet deep just above Stony Creek station, and thence to the north end of the Paleozoic rock valley at Corinth where it turns abruptly to the northeast to flow across the Luzerne mountain ridge. The preglacial Hudson certainly did not flow through the Stony creek gorge, but rather, where the gorge now is, there was an important divide. Among other proofs for this former divide are: the deep, narrow gorge of recent origin; the flaring of the valley both northward and southward from the gorge; and the anomalous turns of both the Hudson and Schroon rivers toward the southwest through a highland region of hard rock, instead of southeastward across the much lower land between Warrensburg and Lake George. The most probable preglacial channel was past Warrensburg, Caldwell, and Glens Falls as shown on the map. The now extinct Luzerne river started on the Stony creek divide, and flowed southward past Corinth and thence through the Paleozoic rock valley to the west of Saratoga Springs. The cause of the passage of the Hudson over the Stony creek divide was partly due to a lowering of the divide by ice erosion, but mostly to the fact that during the ice retreat the ice lobe in the Lake George depression forced the Hudson river to take a more westerly course which was continued after the melting of the ice. The deflection of the river across the Luzerne mountain divide was certainly caused by heavy drift accumulations in the Paleozoic rock valley south of Corinth.

The famous Ausable chasm in Clinton county is a fine illustration of a deep, narrow gorge cut through the Potsdam sandstone by the Ausable river since the Ice age. The river was deflected from its preglacial channel by heavy drift filling.

According to evidence recently presented by Fairchild, the lower portion of the Black river did not flow, as now, westward past Watertown and into the Ontario basin, but continued northward to northeastward into the St Lawrence valley and in perfect

harmony with the rock structures and other drainage lines. The diversion from the preglacial course was due to heavy glacial accumulations between the villages of Black River and Evans Mills.

The deep, narrow gorges which have been cut through the steep eastern and northeastern fronts of the Tug Hill plateau are commonly called "gulfs." Of these the Whetstone gulf (see plate 8) is the most interesting and though little known it is one of the finest examples of its kind in the State. Its length is two miles, and for one mile it shows a depth of 300 feet. The walls are very steep-sided to nearly vertical, especially in the upper end (narrows) where there is just room enough for the swift stream at the bottom. This gulf is certainly postglacial in origin and has been cut into the soft Lorraine and Utica shales. During glacial times the shales were eroded back by the ice (see above) causing the development of the steep eastern front of Tug hill. After the ice disappeared, all east-bound streams from Tug hill, not in their preglacial channels, rushed over the steep shale front and began to erode notches into its summit. These notches were rapidly deepened in the soft shales to form the gulfs whose heads have since been cut back to their present positions.

The world famous Niagara Falls and gorge are wholly postglacial in origin. After plunging 167 feet at the falls, the river rushes for 7 miles through the gorge whose depth is between 200 and 300 feet (plates 47, 48, 49 and 50). When the glacial waters in the eastern Great Lakes region had dropped to the Iroquois level, the Niagara limestone terrace in the vicinity of Buffalo and with steep escarpment or northern front at Lewiston and Queenston, ceased to be covered by lake water, and the Niagara river came into existence by flowing northward over this limestone plain. The river first plunged over the escarpment at Lewiston and Queenston, thus inaugurating Niagara Falls there. Since that time the falls have receded the 7 miles up stream to their present position. In figure 38 we see that soft shales underlie the hard layer of Niagara limestone, and the recession of the falls has clearly been caused by the breaking off of blocks of limestone due to undermining of the soft shales. A glance at the map (plate 50) will show that the gorge development is really taking place on the Horseshoe falls side where the volume of water is much greater, and that in a short time, geologically considered, the American falls will be dry. The rocks exposed in the gorge walls are Niagara limestone, under which in regular order come Niagara (Rochester) shale, Clinton

shale, and Medina shale and sandstone. These formations show only a slight southward dip or tilt.

The Genesee river from its source to Portageville, Wyoming county, appears to be in a mature preglacial valley. Near Portageville, however, the river plunges into a deep, narrow, rock gorge of postglacial origin, which continues for 25 miles to Mount Morris. This gorge has been cut through soft Devonic shales and shaly sandstones, and its walls are mostly nearly vertical, often rising to heights of several hundred feet. The three noted Portage falls (see plate 51) are situated just below Portageville, the upper falls

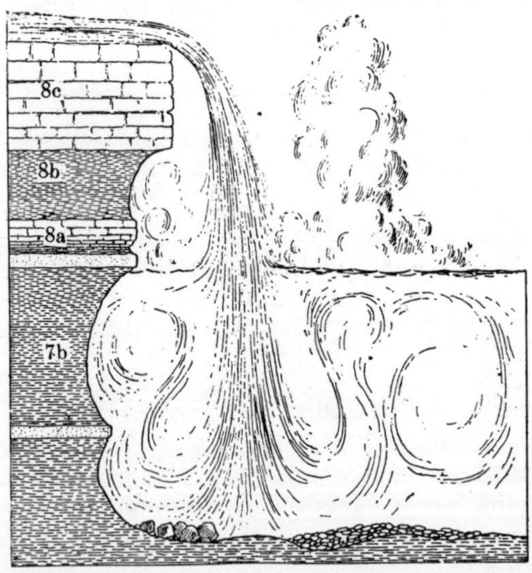


FIG. 38 Section at Niagara Falls, showing the character and position of the rock formations and the depth of water below the falls.
After Gilbert

plunging 66 feet, the middle falls 110 feet, and the lower falls 96 feet. According to Grabau, the preglacial course between Portageville and Mount Morris was farther westward along the present Oatka creek. A second postglacial gorge is entered at Rochester, and this continues for 7 miles to the mouth of the river. Here, also, are three falls, the first over Niagara limestone being 98 feet, the second over Clinton shale and limestone being 20 feet, and the third over Medina sandstone being 105 feet. The preglacial channel here was probably a little to the east and through Irondequoit bay.

In the southern Finger lakes region of south-central New York there are numerous postglacial gorges, a few of the best known ones being: Watkins and Havana glens near the southern end of Seneca lake, Taughannock gorge on the west side of Cayuga lake and in northern Tompkins county, and the gorges of Butternut (Enfield), Fall, Six Mile, and Buttermilk creeks in the vicinity of Ithaca. These gorges all contain waterfalls and have been cut into Devonic shales or sandy shales by streams which have been either partly or wholly diverted from their preglacial courses due to heavy drift filling. In some cases, as at Watkins and Taughannock, the main north-south Seneca and Cayuga valleys were scoured and somewhat deepened by ice erosion, while in all cases the tributary channels were heavily drift filled, thus accounting for the frequent postglacial diversion of these streams which were forced to cut new channels into the steep slopes facing the main valleys.

Watkins glen is several miles long, often very narrow, and with a maximum depth of over 300 feet. Taughannock gorge, which is one and a quarter miles long and with greatest depth of about 350 feet, has in it Taughannock falls whose height is 215 feet and which takes rank as the highest true waterfall in New York State (see frontispiece). Fall creek gorge, on the north side of Cornell campus, is about a mile long and with greatest depth of about 200 feet, and contains Triphammer and Ithaca falls. The Butternut creek (Enfield) gorge is two miles long and with maximum depth of over 300 feet.

In Chautauqua county there are numerous gorges or so-called gulfs which have been cut through the steep front or escarpment of the western border of the Southwestern plateau province. A fine example is the gulf south of Westfield, which is several miles long and from 300 to 400 feet deep. These are also postglacial channels which have been worn into the soft Devonic shales. The steepness of the shale escarpment here, as in the case of Tug hill, was more than likely produced by ice erosion, while the preglacial north-flowing streams had their channels partially or completely filled with glacial debris so that the streams now often flow in postglacial channels. The conditions are here very similar to those of the Finger lakes region already described.

**Length of time since the Ice age.** Estimates of the duration of the glacial epoch by the most able students of the subject vary from 500,000 to 1,500,000 years, these estimates being based on such criteria as amount of erosion and weathering of the earliest till sheets (in Mississippi valley), times necessary for the various

advances and retreats of the ice sheets etc. Although a closer calculation is well nigh impossible because of the variability of the factors involved from time to time, it is nevertheless certain that, from the geological standpoint, the Ice age was of short duration, while, from the standpoint of known human history, it was immensely long.

Estimates of the length of time since the close of the Ice age are perhaps more satisfactory, though it must be remembered that the close of the Ice age was not at the same time for all places. The ice retreated northward very slowly and when, for example, southern New York was free from the ice, northern New York was still glaciated. The best estimates for the length of time since the close of the Ice age in New York State are based upon the rate of recession of Niagara falls. We have learned that the Niagara river began its work about the time the glacial waters in the Erie-Ontario regions had dropped to the Iroquois level, and that the falls were first formed by the plunging of the river over the Niagara limestone escarpment at Queenston and Lewiston. Studies based upon actual surveys, drawings, daguerreotypes, photographs etc. made between the years 1842 and 1905, have shown that the Horseshoe fall had receded about 5 feet a year, while the American fall between 1827 and 1905, had receded about 3 inches a year. Thus the gorge cutting is clearly taking place on the Canadian side. The

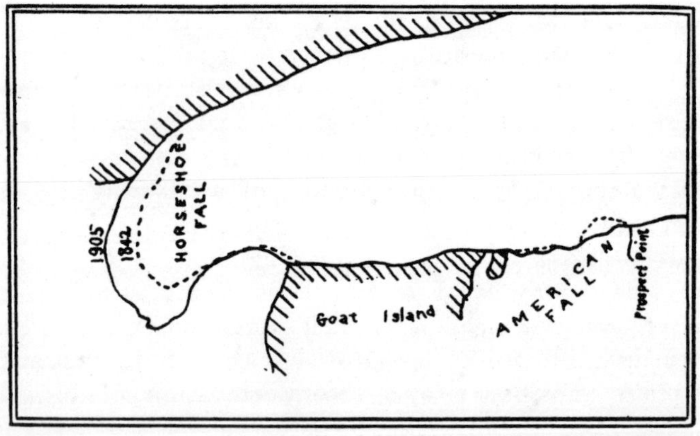


FIG. 39 Sketch map showing the relation of the crest of Niagara Falls in 1842 to that of 1905. Based upon actual surveys. The retreat of the inner portion of the Horseshoe Fall was more than 300 feet.

Modified after Gilbert, U. S. G. S. Bul. 306, p. 20

length of the gorge is 7 miles, and if we consider the rate of recession to have been always 5 feet a year, the length of time necessary to cut Niagara gorge would be something over 7000 years. But the problem is not so simple, since we know that at the time of, or shortly after, the beginning of the river, the upper lakes drained out through the Trent river, and then still later through the Ottawa river. So it is evident that, for a good part of the time since the ice retreated from the Niagara region, the volume of water passing over the falls was notably diminished, and hence the length of time for the gorge cutting increased. The best estimates for the length of time since the ice retreated from the Niagara region vary from 7000 to 50,000 years, an average being about 25,000 years. In a similar way the time based upon the recession of St Anthony's falls, Minnesota, range from about 10,000 to 16,000 years. While closer estimates are practically impossible, it is at least certain that the time since the Ice age is far less than its duration, and that, for the region of New York State, the final ice retreat occurred only a very short time ago.

When we consider the slight amount of weathering and erosion of the latest glacial drift, we are also forced to conclude that the time since the close of the Ice age in New York is to be measured only by some thousands of years. Thus kames, drumlins, extinct lake deltas, and moraines with their kettle holes have generally been very little affected by ice erosion since their formation.

**Champlain subsidence and recent elevation of New York State.** We have already shown that at about the beginning of the glacial epoch the region of New York State, especially along the eastern side, was much higher than it is today, positive proof for this being afforded by the submerged Hudson river channel which must have been cut when the land was higher. Toward the close of the Ice age and shortly after (Champlain epoch), we know that the land had subsided to a level even lower than that of today. It was during this period of subsidence that the lower Hudson and St Lawrence channels were submerged and the sea coast was transferred to more nearly its present position. But as the land was even lower than now, the lowlands of Long island and in the vicinity of New York City were under water and a narrow arm of the sea extended through the Hudson and Champlain valleys to join a broad arm of the sea which reached up the St Lawrence valley and even into the Ontario basin (see figure 34). This Champlain sea existed at the time of the Nipissing Great Lakes already described. Champlain sea beaches, containing marine shells and

the bones of walruses and whales, have been found at altitudes of about 400 feet near the southern end of Lake Champlain, to 500 feet at its northern end, and 600 or more feet at the eastern end of Lake Ontario. In the lower Hudson river valley the deposits of this age are about 70 feet above sea level, and at Albany a little over 300 feet. The altitudes of these so-called raised beaches show how much lower the land was during the time of greatest submergence, and that the subsidence was most toward the north.

The most recent movement of the earth's crust over the area of the State was the very recent gradual elevation which expelled the Champlain sea and left the land at its present altitude. The altitudes of the raised Champlain beaches show that the greatest elevation was on the north. The warping of the Iroquois beaches already described occurred at this same time. Actual surveys during the past century have proved that the upward movement in the northern Great Lakes region is still progressing at the rate of 5 inches in 100 miles in 100 years.

# APPENDIX

## CONSTRUCTION AND USES OF GOVERNMENT CONTOUR MAPS

A number of plates, comprising portions of government topographic (contour) maps, have been introduced into this book for the purpose of illustrating the typical relief features of various parts of the State. Since many persons are not familiar with these maps and their uses, a brief explanation is here given.

These topographic maps, which are called sheets or quadrangles, are rectangular in shape and bounded by latitude and longitude lines. The size of each map is about 17½ inches high by 11½ to 16 inches wide, the latter varying with the latitude. In New York State the scale is nearly always 1 to 62,500 or nearly one mile to the inch, such a sheet or quadrangle covering an area of just one-sixteenth of a square degree. The most valuable feature of these maps is the fact that the surface configuration (relief) of the country is so accurately shown, this feature being explained by the accompanying figures and the following description which is generally found printed on the back of each map: "Relief is shown by contour lines in brown. Each contour passes through points which have the same altitude. One who follows a contour on the ground will go neither up hill nor down hill, but on a level. By the use of the contours not only are the shapes of the plains, hills, and mountains shown, but also the elevations. The line of the sea coast itself is a contour line, the datum or zero of elevation being the mean sea level. The contour line at, say, 20 feet above sea level is the line that would be the sea coast if the sea were to rise or the land to sink 20 feet. Such a line runs back into the valleys and forward around the points of hills and spurs. On a gentle slope this contour line is far from the present coast line, while on a steep slope it is near it. Thus a succession of these contour lines far apart on the map indicates a gentle slope; if close together, a steep slope; and if the contours run together in one line, as if each were vertically under the one above it, they indicate a cliff. . . . The contour interval, or vertical distance in feet between one contour and the next, is stated at the bottom of each map. This interval varies according to the character of the area mapped; in a flat country it may be as small as 5 feet; in a mountainous region it may be 200 feet. Certain contours, usually every fifth one, are

accompanied by figures stating elevation above sea level. The heights of many definite points, such as road corners, railroad stations, railroad crossings, summits, water surfaces, triangulation stations, and bench marks, are also given. The figures in each case are placed close to the point to which they apply, and express the elevation to the nearest foot only. . . . All water features are shown in blue, the smaller streams and canals in full blue lines, and the larger streams, lakes, and the sea by blue water lining. . . . The works of man are shown in black, in which coloring all letter-

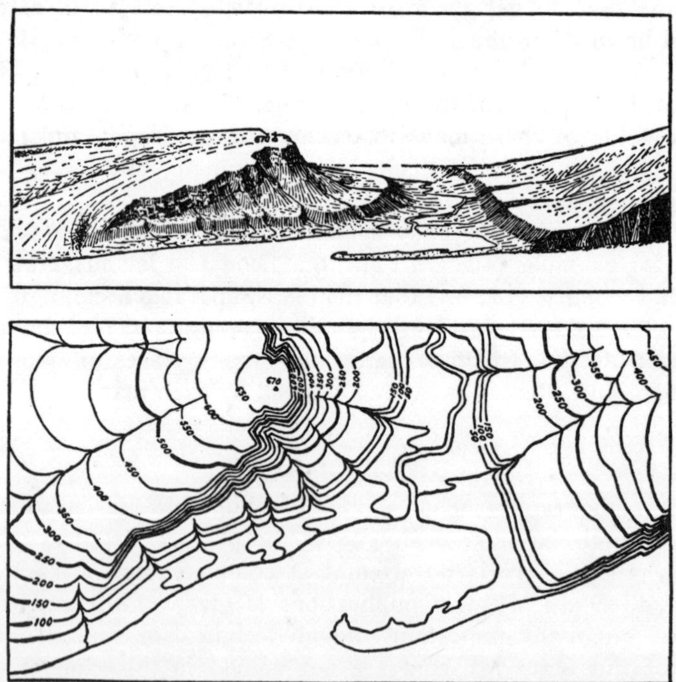

FIG. 40 Ideal sketch and corresponding contour map (U. S. G. S.).

ing also is printed. . . . Houses are shown by small black squares which in the densely built portions of cities and towns merge into blocks. Roads are shown by fine double lines, trails by single dotted lines, and railroads by full black lines with cross lines. Other cultural features are represented by conventions which are easily understood. The sheets composing the topographic atlas are designated by the name of a principal town or of some prominent natural feature within the quadrangle and the names of the adjoining published sheets are printed on the margins. They are sold

at ten cents each when fewer than 50 copies are purchased, but when ordered in lots of 50 or more copies, whether of the same or of different sheets, the price is six cents each."

These maps are published by the United States Geological Survey, and orders for them should be sent to the director of that bureau at Washington, D. C. The order should be accompanied by cash or a post office money order. Each quadrangle has a special name by which it must be ordered. A large portion of New York State has been covered by such topographic surveys and, in order to know how to get the map covering a given region, reference should be made to the Index to Atlas Sheets for New York State. This index may be procured free of charge by dropping a post card to the Director of the United States Geological Survey.

The value of these maps to teachers of geography and physical geography would be difficult to overestimate, and every school should have a supply of these maps readily accessible. For the teaching of home geography as well as that of other parts of the State, for example, Niagara Falls, the Thousand Islands, New York City and vicinity, etc., no other map is comparable because, in addition to the ordinary features, the relief (topography) of the land is shown in detail. In other states, also, many places of importance or geographic interest have been covered by such maps.

## BIBLIOGRAPHY

This bibliography includes only the bulletins and papers of a more general character dealing with the physical features of New York State. There is no attempt at completeness. An extended list of all State Museum publications is given at the end of this bulletin, but many important, though technical or special, publications of the State Museum are not named in this chapter. The most exhaustive lists of papers dealing with the geology of New York, both those issued by the New York State Museum and published elsewhere, are to be found in United States Geological Survey Bulletins 127, 188, 189, 301, 372, 409, 444 and 495, covering the years 1732 to 1910 inclusive. For still later years other bulletins will appear. By referring to "New York" in the index, the subjects and regions treated may be readily found. Numerous references are also given in Tarr's Physical Geography of New York State. The reader who desires to know what scientific publications of the New York survey refer to a given subject or region should address the Director of the State Museum, Education Building, Albany, N. Y.

## New York State Museum Bulletins

19 **Merrill, F. J. H.** Guide to the Study of the Geological Collections of the New York State Museum. 1898

21 **Kemp, J. F.** Geology of the Lake Placid Region. 1898

34 **Cumings, E. R. & Prosser, C. S.** Lower Silurian System of Eastern Montgomery County and Stratigraphy of the Mohawk Valley. 1900

42 **Ruedemann, R.** Hudson River Beds near Albany. 1901

45 **Grabau, A. W.** Geology of Niagara Falls and Vicinity. 1901

48 **Woodworth, J. B.** Pleistocene Geology of Nassau County and Borough of Queens. 1901

63 **Clarke, J. M. & Luther, D. D.** Stratigraphy of Canandaigua and Naples Quadrangles. 1904

77 **Cushing, H. P.** Geology of the Vicinity of Little Falls. 1905

81 **Clarke, J. M. & Luther, D. D.** Watkins and Elmira Quadrangles. 1905

82 **Clarke, J. M.** Geologic Map of the Tully Quadrangle. 1905

83 **Woodworth, J. B.** Pleistocene Geology of the Mooers Quadrangle. 1905

84 —— Ancient Water Levels of the Champlain and Hudson Valleys. 1905

92 **Grabau, A. W.** Geology and Paleontology of the Schoharie Region. 1906

95 **Cushing, H. P.** Geology of the Northern Adirondack Region. 1905

96 **Ogilvie, I. H.** Geology of the Paradox Lake Quadrangle. 1905

99 **Luther, D. D.** Geology of the Buffalo Quadrangle. 1906

101 —— Geology of the Penn Yan-Hammondsport Quadrangles. 1906

106 **Fairchild, H. L.** Glacial Waters in the Erie Basin. 1907

111 —— Drumlins of New York. 1907

114 **Hartnagel, C. A.** Geological Map of the Rochester and Ontario Beach Quadrangles. 1907

115 **Cushing, H. P.** Geology of the Long Lake Quadrangle. 1907

118 **Clarke, J. M. & Luther, D. D.** Geologic Maps and Descriptions of the Portage and Nunda Quadrangles including a Map of Letchworth Park. 1908

126 **Miller, W. J.** Geology of the Remsen Quadrangle including Trenton Falls and vicinity. 1909

127 **Fairchild, H. L.** Glacial Waters in Central New York. 1909

128 **Luther, D. D.** Geology of the Geneva-Ovid Quadrangles. 1909

135 **Miller, W. J.** Geology of the Port Leyden Quadrangle. 1910

137 **Luther, D. D.** Geology of the Auburn-Genoa Quadrangles 1910

138 **Kemp, J. F. & Ruedemann, R.** Geology of the Elizabethtown and Port Henry Quadrangles. 1910

145 **Cushing, H. P.; Fairchild, H. L.; Ruedemann, R., & Smyth, C. H.** Geology of the Thousand Islands Region. 1910

146 **Berkey, C. P.** Geologic Features and Problems of the New York City Aqueduct. 1911

148 **Gordon, C. E.** Geology of the Poughkeepsie Quadrangle 1911

152 **Luther, D. D.** Geology of the Honeoye-Wayland Quadrangles. 1911

153 **Miller, W. J.** Geology of the Broadalbin Quadrangle. 1911

154 **Stoller, J. H.** Glacial Geology of the Schenectady Quadrangle. 1911

160 **Fairchild, H. L.** Glacial Waters in the Black and Mohawk Valleys. 1912

169 **Cushing, H. P. & Ruedemann, R.** Geology of Saratoga Springs and Vicinity.

**Hopkins, T. C.** Geology of the Syracuse Quadrangle.

**Miller, W. J.** Geology of the North Creek Quadrangle.

**Natural History Survey of New York. Division 4 (Geology). 1842-43**

For this survey the State was divided into four districts and the reports cover all the counties of the State.

V. 1 pt 1 **Mather, W. W.** First Geological District. 1843

V. 2 pt 2 **Emmons, E.** Second Geological District. 1842

V. 3 pt 3 **Vanuxem, L.** Third Geological District. 1842

V. 4 pt 4 **Hall, J.** Fourth Geological District. 1843

**New York State Museum handbooks**

No. 15 **Clarke, J. M.** Guide to Excursions in the Fossiliferous Rocks of New York. 1899

No. 19 **Hartnagel, C. A.** Classification of the Geologic formations of the State of New York. 1912

**New York State Museum State maps**

Economic and Geologic Map of the State of New York. Scale 14 miles to 1 inch. 1894

Geologic Map of New York. Scale 5 miles to 1 inch. 1901

Map of New York Showing the Surface Configuration and Water Sheds. 1901

**United States Geological Survey folios**

No. 83 New York City and Vicinity. By Merrill, Darton, Hollick, Willis, Salisbury, Dodge, & Pressey.

No. 169 Watkins Glen-Catatonk. By Williams, Tarr & Kindle.

**Miscellaneous papers**

**Baldwin, S. P.** Pleistocene History of the Champlain Valley. Amer. Geologist, 13: 170-84. 1894

**Bishop, I. P.** Structural and Economic Geology of Erie County. Fifteenth Ann. Rep't N. Y. State Geol., p. 305-92. 1897

**Brigham, A. P.** The Geology of Oneida County. Oneida Nat. Hist. Soc. Trans. for 1888, p. 102-18

—— The Finger Lakes of New York. Amer. Geo. Soc. Bul. 25, p. 203-23. 1893

—— Glacial Flood Deposits in the Chenango Valley. Geol. Soc. Amer. Bul. 8, p. 17-30. 1897

—— Topography and Glacial Deposits of the Mohawk Valley. Geol. Soc. Amer. Bul. 9, p. 183-210. 1898

—— Glacial Geology of the Broadalbin, Gloversville, Amsterdam and Fonda Quadrangles. N. Y. State Mus. Bul. 121, p. 21-31. 1908

**Berkey, C. P.** Structural and Stratigraphic Features of the Basal Gneisses of the Highlands. N. Y. State Mus. Bul. 107, p. 361-78. 1906

—— Areal and Structural Geology of Southern Manhattan Island. N. Y. Acad. Sci., 19: 247-82. 1909

**Clarke, J. M.** Brief Outline of the Geological Succession in Ontario County. N. Y. State Geol. Rep't 4, p. 9-22. 1885

**Clarke, J. M.** Report of Fieldwork in Chenango County. 47th Ann. Rep't N. Y. State Mus., p. 725-51. 1894

**Cushing, H. P.** Report on the Geology of Clinton County. N. Y. State Geol. Rep't 13, p. 473-89. 1894

—— Report on the Geology of Clinton County. N. Y. State Geol. Rep't 15, p. 499-573. 1895

—— Report on the Geology of Franklin County. N. Y. State Geol. Rep't 18, p. 73-128. 1899

—— Geology of Rand Hill and Vicinity, Clinton County. N. Y. State Geol. Rep't 19, p. 37-82. 1901

—— Geologic Work in Franklin and St Lawrence Counties. N. Y. State Geol. Rep't 20, p. 23-95. 1902

**Darton, N. H.** Report on the Geology of Albany County. N. Y. State Geol. Rep't 13, p. 229-61. 1894

—— Report on the Geology of Ulster County. N. Y. State Geol. Rep't 13, p. 289-372. 1894

—— Geology of the Mohawk Valley in Herkimer, Fulton, Montgomery and Saratoga Counties. N. Y. State Geol. Rep't 13, p. 407-29. 1894

—— Shawangunk Mountain. Nat. Geog. Mag. 6: 23-34. 1894

—— Description of the Faulted Region of Herkimer, Fulton, Montgomery and Saratoga Counties. N. Y. State Geol. Rep't 14, p. 31-53. 1895

**Fairchild, H. L.** Pleistocene Geology of Western New York. N. Y. State Geol. Rep't 20, p. r103-39. 1902

—— Latest and Lowest Pre-Iroquois Channels between Syracuse and Rome. N. Y. State Geol. Rep't 21. 1903

—— Glacial Waters from Oneida to Little Falls. N. Y. State Geol. Rep't 22, p. r17-41. 1904

**Gilbert, G. K.** Niagara Falls and their History. Nat. Geog. Monograph v. 1, no. 7, p. 203-36. 1895

—— Rate of Recession of Niagara Falls. U. S. G. S. Bul. 306. 1907

**Grabau, A. W.** The Preglacial Channel of the Genesee River. Boston Soc. Nat. Hist. Proc., 26: 359-69. 1894

**Gratacap, L. P.** Geology of the City of New York. Third ed. Henry Holt Co. 1909

**Hall, J. & Clarke, J. M.** Stratigraphic and Faunal Relations of the Oneonta Sandstones, the Ithaca and the Portage Groups in Central New York. With Maps. N. Y. State Geol. Rep't 15, p. 27-81. 1897

**Hartnagel, C. A.** Formations of the Skunnemunk Mountain Region. N. Y. State Mus. Bul. 107, p. 39-54. 1906

**Heilprin, A.** The Catskill Mountains. Amer. Geog. Soc. Bul. 39, no. 4, p. 193-99. 1907

**Hobbs, W. H.** Origin of the Channels Surrounding Manhattan Island. Geol. Soc. Amer. Bul. 16, p. 151-82. 1905

**Kemp, J. F.** Report on the Geology of Essex County. N. Y. State Geol. Rep't 13, p. 431-72. 1894; and N. Y. State Geol. Rep't 15, p. 575-614. 1897

—— Physiography of the Eastern Adirondack Region in the Cambrian and Ordovician Periods. Geol. Soc. Amer. Bul. 8, p. 408-12. 1897

—— **& Newland, D. H.** Report on the Geology of Washington, Warren and Parts of Essex and Hamilton Counties. N. Y. State Geol. Rep't 17, p. 499-553. 1899

—— **Newland, D. H. & Hill, B. F.** Report on the Geology of Hamilton, Warren, and Washington Counties. N. Y. State Geol. Rep't 18, p. 137-62. 1899

—— **& Hill, B. F.** Report on the Precambrian Formations in Parts of Warren, Saratoga, Fulton and Montgomery Counties. N. Y. State Geol. Rep't 19, p. 117-35. 1901

—— Physiography of the Adirondacks. Popular Sci. Monthly, 68: 195-210. March 1906

**Kümmel, H. B.** The Newark or New Red Sandstone Rocks of Rockland County. N. Y. State Geol. Rep't 18, p. 9-50. 1899

**Lincoln, D. F.** Structural and Economic Geology of Seneca County. N. Y. State Geol. Rep't 14, p. 57-125. 1895

**Luther, D. D.** Economic Geology of Onondaga County. N. Y. State Geol. Rep't 15, p. 237-303. 1897

—— Brine Springs and Salt Wells of New York, and the Geology of the Salt District. N. Y. State Geol. Rep't 16, p. 171-226. 1899

**Merrill, F. J. H.** Quaternary Geology of the Hudson River Valley. N. Y. State Geol. Rep't 10, p. 103-55. 1891

**Miller, W. J.** Ice Movement and Erosion along the Southwestern Adirondacks. Amer. Jour. Sci., 4th ser., 27: 289-98. 1909

—— Exfoliation Domes in Warren County. N. Y. State Mus. Bul. 149, p. 187-94. 1911.

—— Preglacial Course of the Upper Hudson River. Geol. Soc. Amer. Bul. 22, p. 177-86. 1911

—— Early Paleozoic Physiography of the Southern Adirondacks. To appear in the 9th annual report of the Director of Science Division for 1912

**Prosser, C. S.** Hamilton and Chemung Series of Central and Eastern New York. Pt 1, N. Y. State Geol. Rep't 15, p. 83-222, 1897; pt 22, N. Y. State Geol. Rep't 17, p. 65-315. 1899

—— **& Cumings, E. R.** Lower Silurian (Ordovician) Formations on West Canada Creek and in the Mohawk Valley. N. Y. State Geol. Rep't 15, p. 615-59. 1897

—— **& Rowe, R. B.** Geology of the Eastern Helderbergs. N. Y. State Geol. Rep't 17, p. 329-54. 1897

**Randall, F. A.** Report on the Geology of Cattaraugus and Chautauqua Counties. N. Y. State Geol. Rep't 13, p. 517-27. 1894

**Ries, H.** Geology of Orange County. N. Y. State Geol. Rep't 15, p. 393-475. 1897

**Smyth, C. H.** General and Economic Geology of Four Townships in St Lawrence and Jefferson Counties. N. Y. State Geol. Rep't 13, p. 491-515. 1894

—— Report on the Crystalline Rocks of St Lawrence County. N. Y. State Geol. Rep't 15, p. 477-97. 1897

—— Report on Crystalline Rocks of the Western Adirondack Region. N. Y. State Geol. Rep't 17, p. 469-97. 1899.

—— Geology of the Crystalline Rocks in the Vicinity of the St Lawrence River. N. Y. State Geol. Rep't 19, p. r83-104. 1901

**Tarr, R. S.** Hanging Valleys in the Finger Lakes Region of Central New York. Amer. Geol., 33: 271-91. 1904

**Watson, T. L.** Some Higher Levels in the Postglacial Development of the Finger Lakes of New York State. N. Y. State Mus. Rep't 51 (1), p. r55-117. 1899

**Veatch, A. C.** Outlines of the Geology of Long Island. U. S. G. S., Professional Paper 44, p. 53-85. 1906

# INDEX

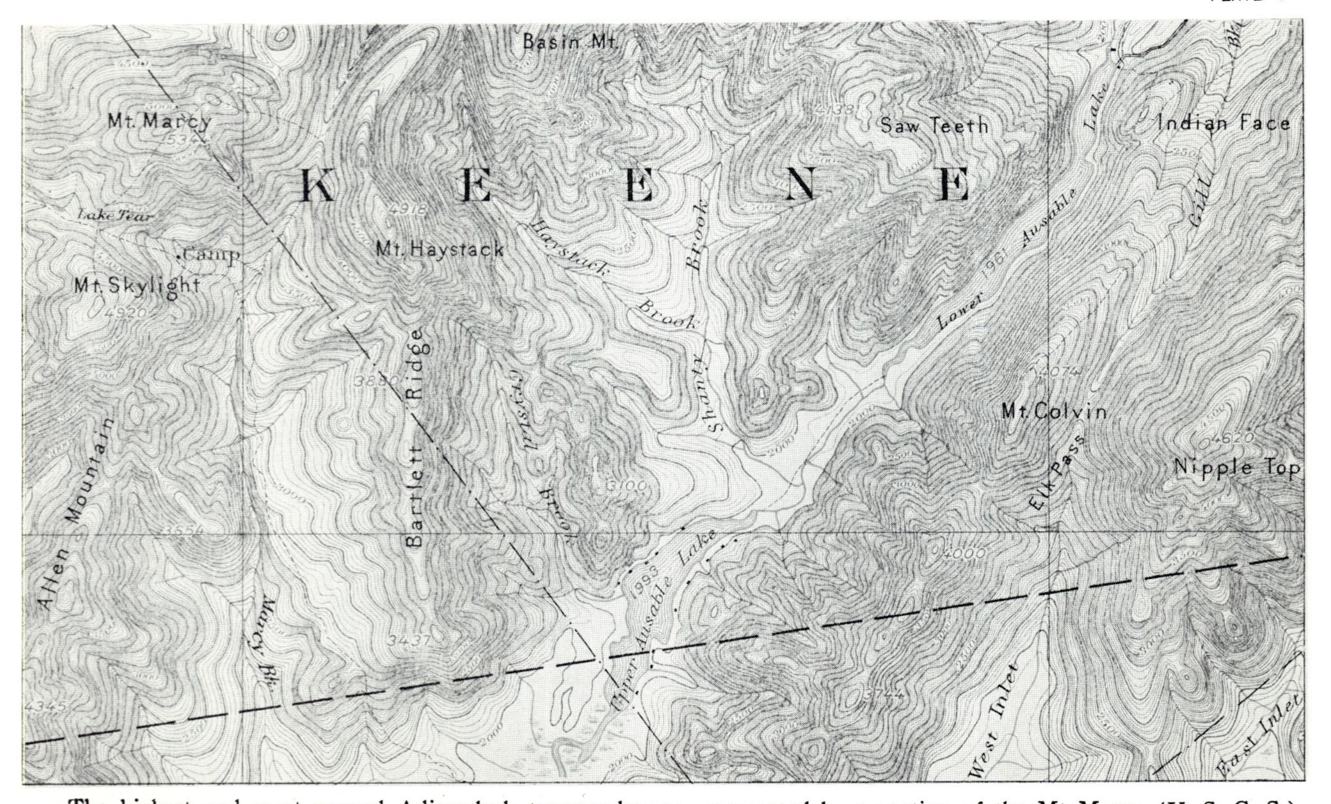


The highest and most rugged Adirondack topography, as represented by a portion of the Mt Marcy (U. S. G. S.) quadrangle. One of the northeast-southwest ridges so common in the eastern Adirondacks is shown just east of the lakes. The rocks are nearly all of plutonic igneous origin (anorthosite). Scale, about 1 mile to the inch. Those who are not familiar with the topographic map and its interpretation should refer to the explanation in the appendix.

## Plate 2

A typical view in the midst of the Adirondacks. Looking across Long lake from Buck mountain, showing Mt Kempshall and Blueberry mountain. Note the flowing outlines of the topography.

From N. Y. State Mus. Bul. 115, pl. 15

## Plate 3

General view in the Adirondacks looking northward across Lake Pleasant (Hamilton county). Note the flowing outlines of the topography.

**Photo loaned by John A. Cole, Lake Pleasant, N. Y**

PLATE 4

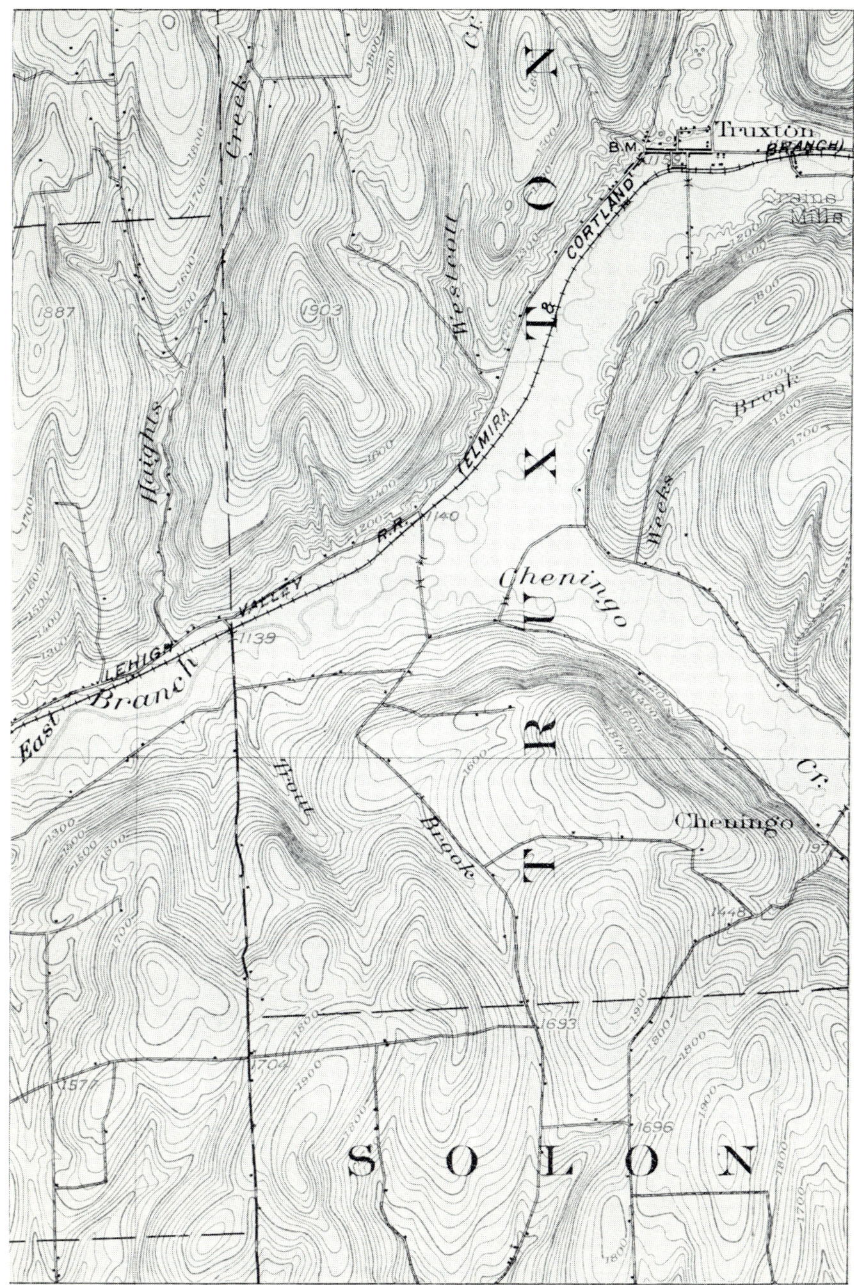


Typical southwestern plateau topography. Note the numerous irregular shaped hills, the high general level of the country and the deep broad-bottomed valleys occupied by the larger streams. The rocks are shales and sandstones of Devonic age. From Cortland (U. S. G. S.) quadrangle. Scale, about 1 mile to the inch.

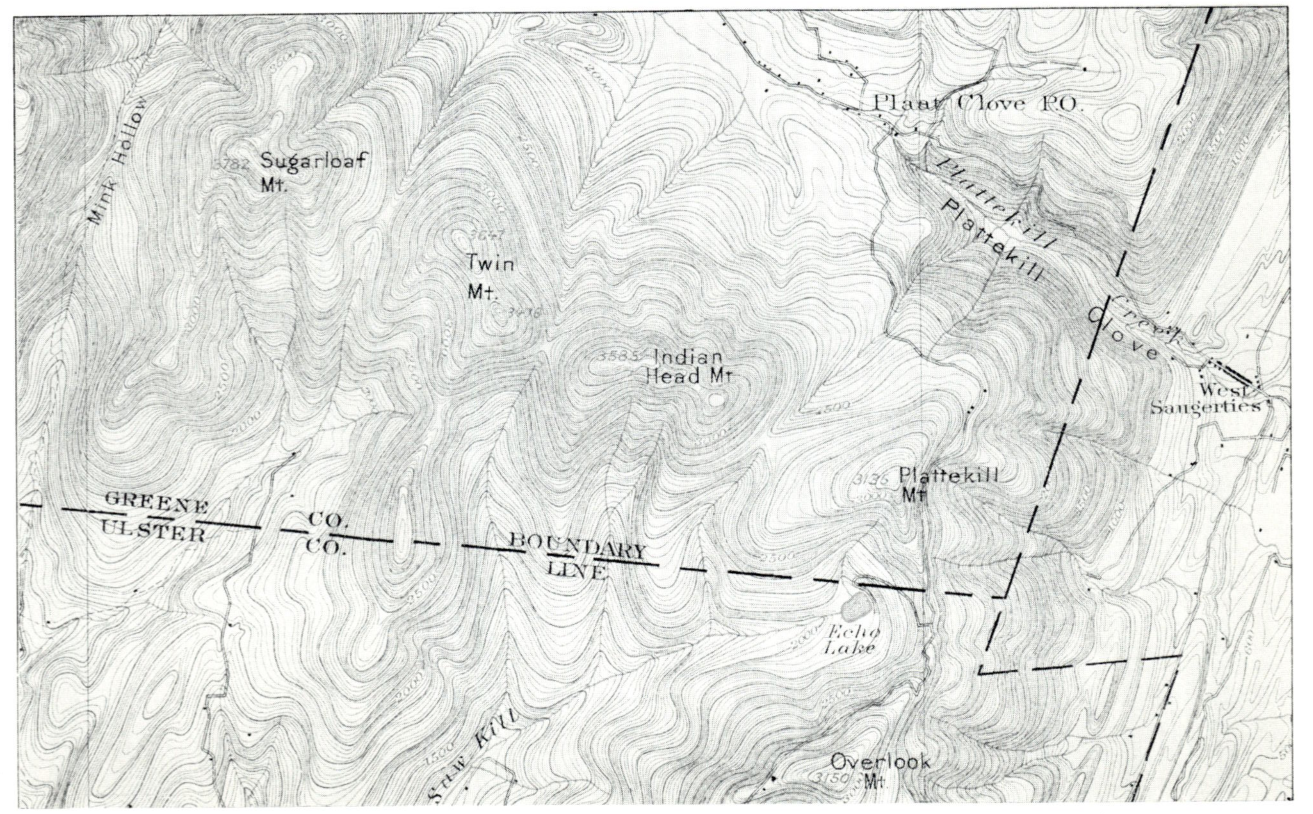


Map of part of the Catskill province showing the high, rugged character of the mountains and the remarkably steep, high eastern front of the province overlooking the great Hudson valley. From the Kaaterskill (U. S. G. S.) quadrangle. Scale, about 1 mile to the inch.

## Plate 6

The pass between Hunter and Plateau mountains in the Catskills. This is a fine example of the deep, steep-sided, V-shaped valleys so characteristic of this province.

From Annual Rep't N. Y. State Geol., 1897, pl. 20, facing p. 282

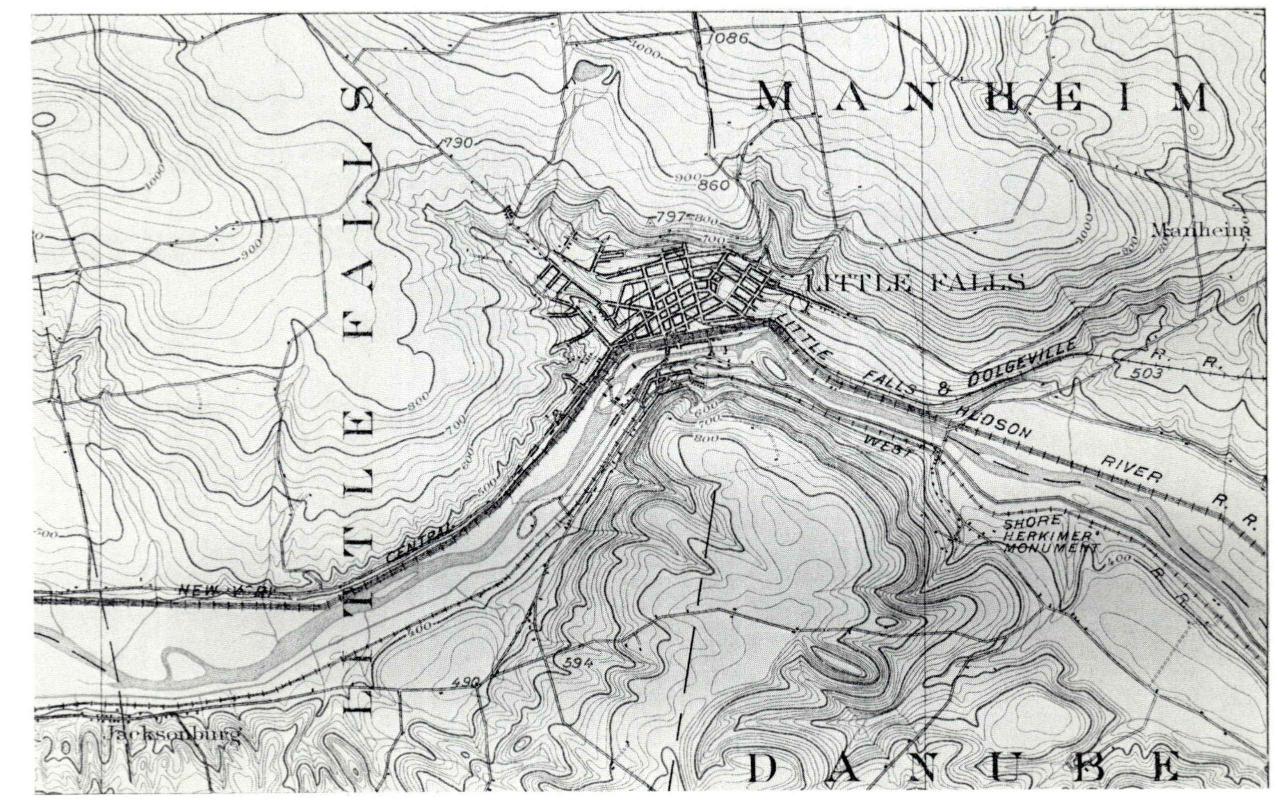


Topographic map showing the deep, narrow gorge of the Mohawk river at Little Falls. The walls of the gorge rise fully 500 feet above the river which is less than 400 feet above sea level. The gorge is postglacial in origin and before the great ice age it was replaced by an important divide with one stream (Mohawk river) flowing eastward and another stream (Rome river) flowing westward. Part of Little Falls (U. S. G. S.) quadrangle. Scale, about 1 mile to the inch.

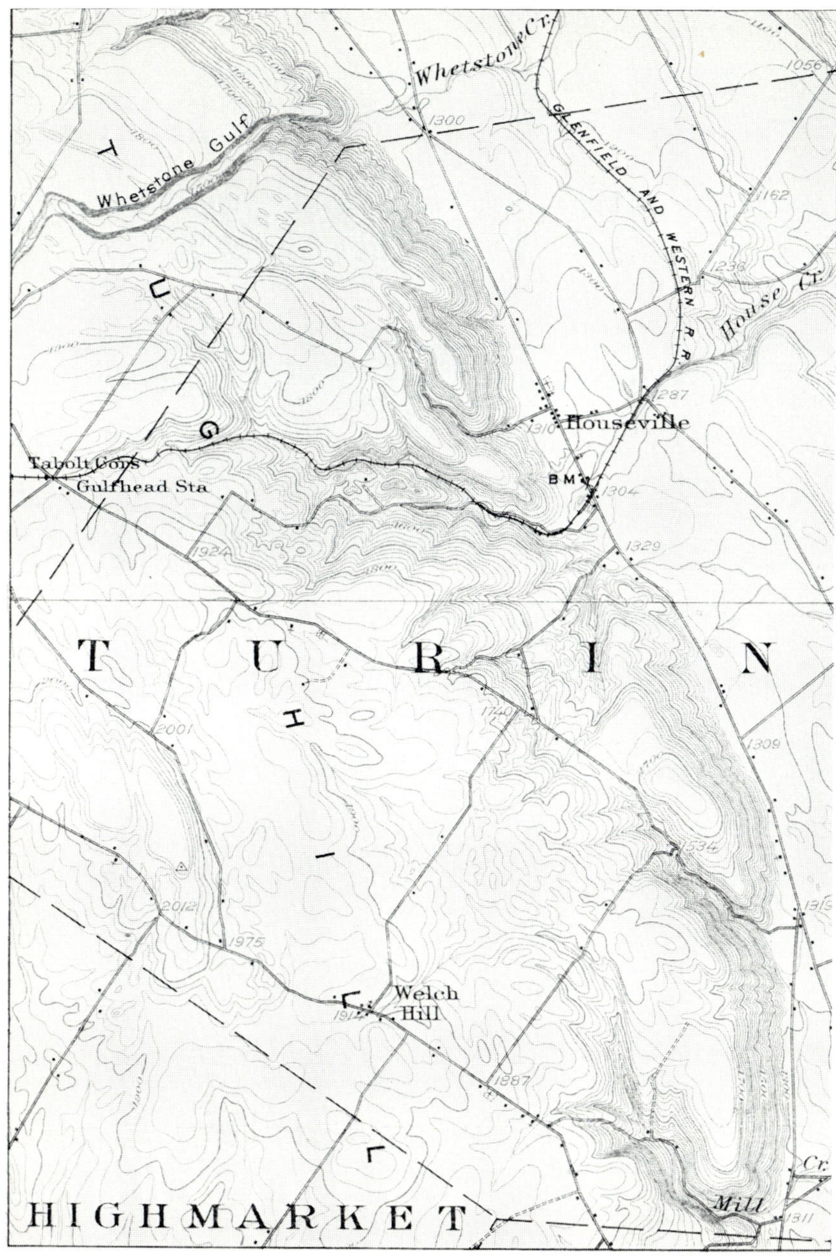


A portion of the Port Leyden (U. S. G. S.) quadrangle, illustrating the topography of the Tug hill plateau. Note the high, general plateau level at from 1800 to 2000 feet above sea level and the abrupt termination of this plateau on the east side. Whetstone gulf is a fine example of numerous gorges cut through the steep plateau front. Scale, about 1 mile to the inch.

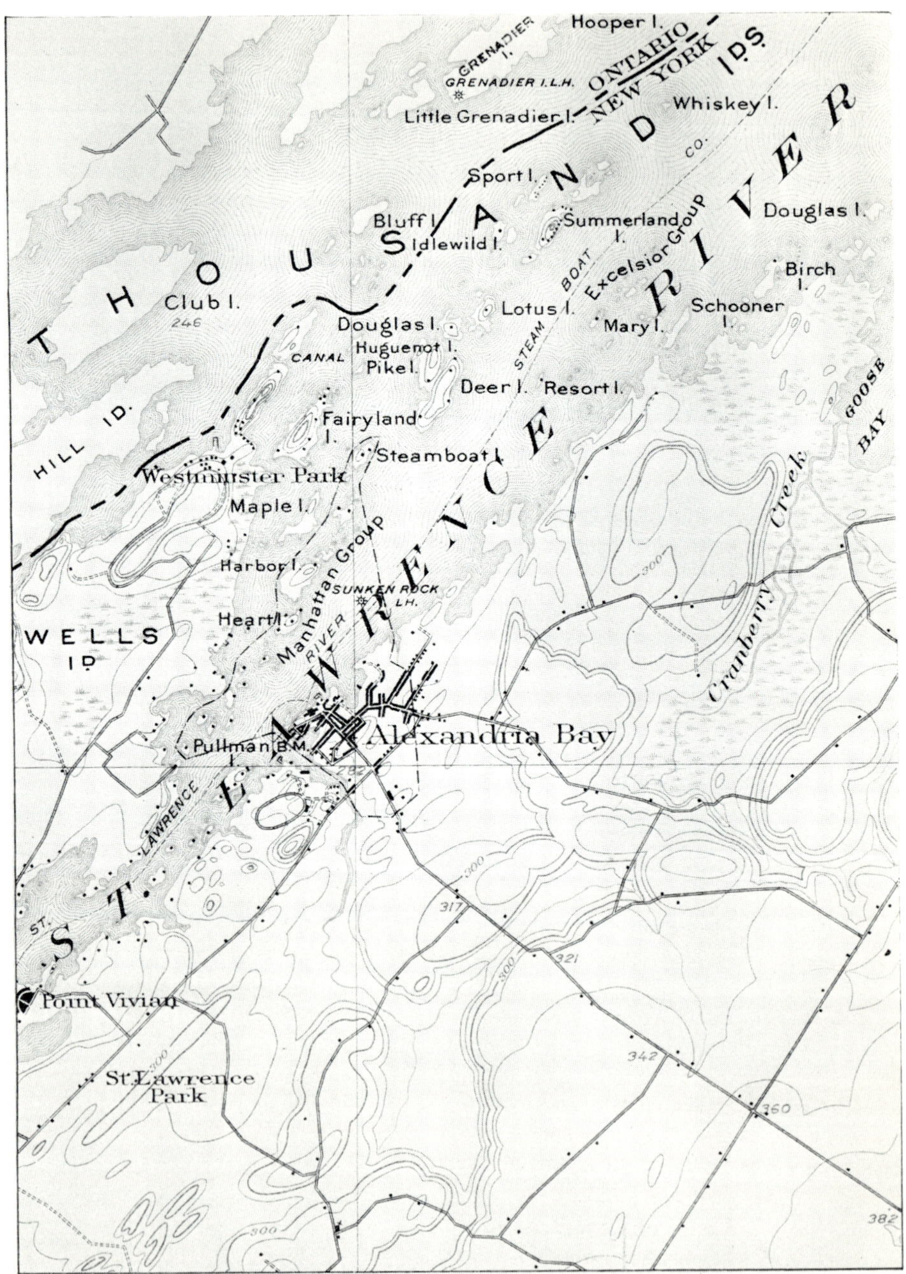


The Thousand Islands region in the vicinity of Alexandria Bay. The broad river, dotted with islands, does not here occupy a distinct stream channel in the usual sense of the term. A removal of the water would show that the topography of the river bottom is in no way essentially different from that of the country just south of Alexandria Bay. From Alexandria Bay (U. S. G. S.) quadrangle. Scale, about 1 mile to the inch.

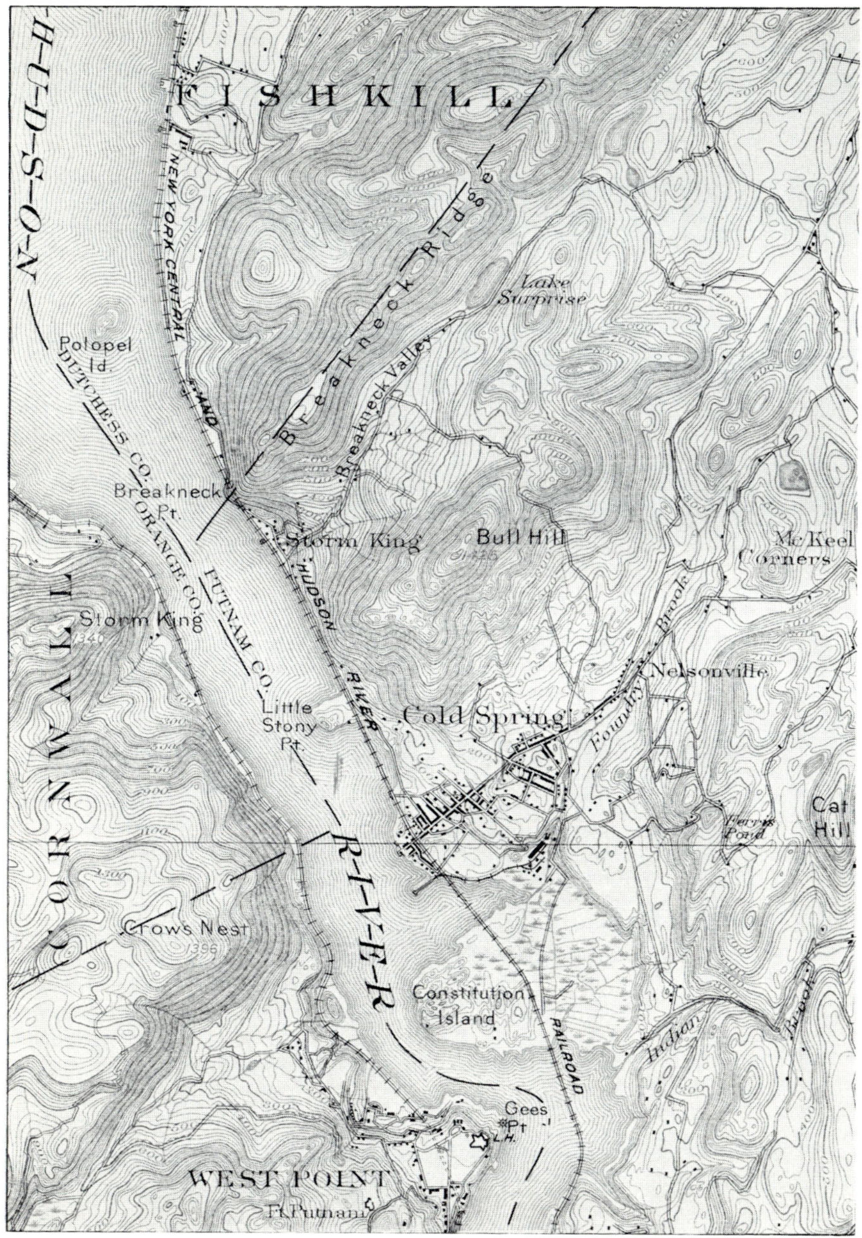


A portion of the West Point (U. S. G. S.) contour map, illustrating the topography of the Highlands of the Hudson. The river here flows through a deep, narrow gorge which has been cut through the northeast-southwest ridges of very hard Precambric rock. The sides of the gorge rise abruptly from the river to heights of 1200 or 1400 feet, while the rock bottom of the gorge is hundreds of feet below the river level. Scale, about 1 mile to the inch.

## Plate 11

Looking north up the Hudson where the river flows through a deep narrow gorge in the hard granitic (Precambric) rocks of the Highlands. The mountain on the left is Crow's Nest and the nearer one on the right is Bull hill. Each of these mountains rises to more than 1400 feet above the river while the rock bottom of the river is buried some 600 to 800 feet. Compare with plate 20.

From N. Y. State Mus. Bul. 146 p,l. 16

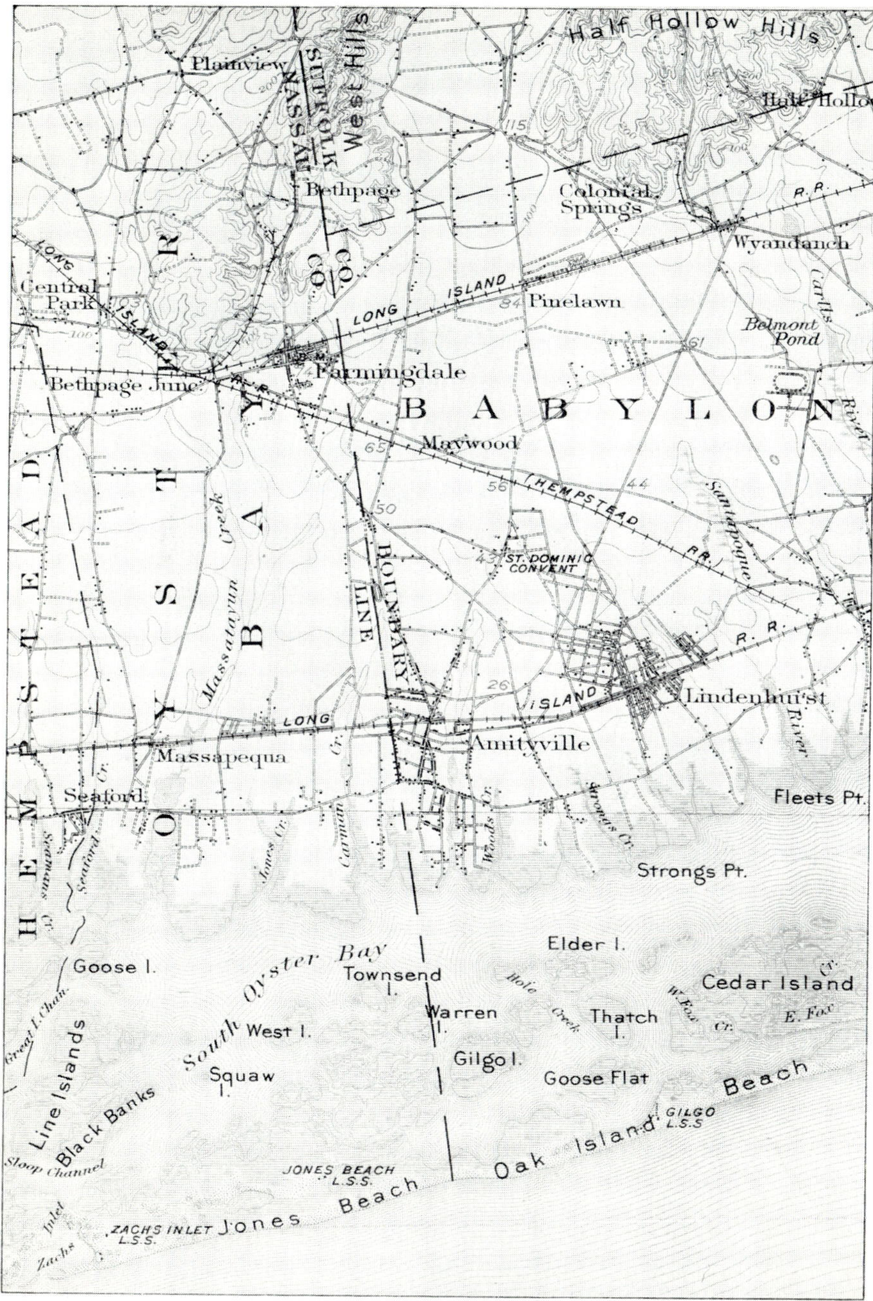


A portion of the Islip (U. S. G. S.) quadrangle showing typical Long Island topography. Note the distinct hills (of glacial morainic origin) on the north, the perfect plain only very slightly stream-dissected and with very gentle seaward slope, and the long beach (built by wave action) inclosing the shallow bay with its many swampy islands. Scale, about 2 miles to the inch.

# Plate 13

An outcrop of typical Grenville (Precambric) limestone showing the folded or contorted and streaked character of the rock. View taken 1½ miles southeast of Johnsburg, Warren county.

W. J. Miller, photo

Plate 14

Upper figure. Typical outcrop of Adirondack Grenville gneiss, showing distinct stratification and tilted beds. Three-quarters of a mile north-northeast of Batchellerville, Saratoga county.

From N. Y. State Mus. Bul. 153, pl. 1

Lower figure. Typical outcrop of Adirondack syenite (igneous rock) at Harrisville, Lewis county.

From N. Y. State Geol. Rep't 1897, facing p. 470

Plate 15

A ledge of Precambric rock near Loon lake in the northern Adirondacks, showing the very ancient Grenville gneiss in sharp contact with syenite. The light and dark banded rock on the right is Grenville and the hammer is on the contact.

From N. Y. State Mus. Bul. 5, pl. 3

## Plate 16

View of Breakneck mountain where the Hudson river cuts across the Highlands. Seen from the shore opposite Cold Spring, Putnam county. The rock is Precambric granite. Back from the river, in these mountains, the character of the topography is much like that of the Adirondacks.

## Plate 17

The upper view is of Potash mountain, locally called the "Potash Kettle," three miles north of Luzerne, Warren county. It is a remarkable feature of the landscape, the almost isolated mountain towering 1100 feet above the surrounding valleys. The rock is Precambric granite.

The lower view shows several dark lava (diabase) dikes cutting gray syenite three-fourths of a mile northwest of Northville, Fulton county. Both rocks are of Precambric age. Photos by W. J. Miller

Plate 18

The Grand Flume of Ausable chasm, Clinton county. The rock is Potsdam sandstone in horizontal layers and the gorge is postglacial.

From N. Y. State Mus. Bul. 19, pl. 22

Plate 19

Potsdam (Upper Cambric) sandstone resting upon the eroded surface of Precambric rock (gneiss) at Corinth, Saratoga county. The difference in age of the two rock masses here in contact is millions of years.

From N. Y. State Mus. Bul. 19, pl. 20

Plate 20

Little Falls (late Cambric) dolomitic limestone resting upon the eroded surface of Precambric rock as seen in a West Shore Railroad cut 1 mile west of Randall, Montgomery county.

From N. Y. State Mus. Bul. 19 pl. 25

Plate 21

Trenton (Ordovicic) limestone at Sherman fall in the gorge at Trenton Falls, Oneida county. The thin-bedded and perfectly stratified character of this impure limestone is here well shown. This famous formation, loaded with fossils, was named from this locality.

From N. Y. State Mus. Bul. 95, pl. 9

Plate 22

Falls over Canajoharie (Trenton) black shale south of Canajoharie, Montgomery county.

From N. Y. State Mus. Bul. 19, pl. 35

Plate 23

A small fold in the Hudson river (Ordovicic) shales and sandstones along the lower part of Catskill creek in Greene county. This fold was produced at the time of the Taconic revolution.

From N. Y. State Mus. Bull. 19, p. 36

## Plate 24

Granite dike in Hudson River or Manhattan schist, as seen on the south side of 192d street, New York City. The schist is metamorphosed Hudson River (Ordovicic) shale and sandstone and the granite, while molten, broke its way through. **The steep tilt of the schist layers was produced by the Taconic disturbance.**

From N. Y. State Mus. Bul. 19, pl. 1

Plate 25

Shawangunk (Siluric) conglomerate resting upon the eroded edges of Hudson River (Ordovicic) shales, thus showing a sharp contact between rocks of two great periods of earth history. The shale layers are more steeply inclined because they were affected by both the Taconic and Appalachian disturbances, while the conglomerate was affected only by the latter disturbance.

From N. Y. State Mus. Rep't 60 (2), 1906, pl. A, facing p. 298

Plate 26

Beach markings on Medina sandstone

Seaweed, Arthrophycus harlani, on Medina sandstone
From N. Y. State Mus. Bul. 19, pl. 45

## Plate 27

Niagara river gorge below the Suspension bridge, as viewed from the Canadian side. Very distinctly stratified beds of lower Siluric (Medina, Clinton and Niagara) age are shown.

From N. Y. State Mus. Bul. 19, pl. 55

Plate 28

The eastern face of Shawangunk mountain, 2 miles south of Lake Mohonk, Ulster county. The light colored rock forming the summit of the ridge is the very resistant Shawangunk conglomerate which protects the soft underlying mass of Hudson River (Ordovicic) shales.

From N. Y. State Mus. Bul. 19, pl. 40

Plate 29

Awosting falls over Shawangunk conglomerate, Peterkill, near Lake Minnewaska, Ulster county

From N. Y. State Mus. Bul. 19, pl. 42

Plate 30

Cliff of Lower Devonic (Coeymans) limestone near Indian Ladder, Albany county

From N. Y. State Mus. Bul. 19, pl. 70

Plate 31

Marcellus and Hamilton (Devonic) shales outcropping along the shore of Lake Erie at Athol Springs, Erie county. The rock is dark colored and in thin layers. The smooth, vertical walls are due to a breaking off of rock masses along what are called "joint planes" and which are so common in nearly all rock ledges.

From N. Y. State Mus. Bul. 19, pl. 72

Plate 32

Upper Devonic (Cashaqua and Rhinestreet) shales in the gorge of the Genesee river near Mount Morris. The thin bedded and stratified character of these rocks are well shown.

Plate 33

Hydraulic limestone or waterlime of Salina (Siluric) age at the cement quarries 1 mile south of Whiteport, Ulster county. The pronounced tilt of the strata was given at the time of the Appalachian revolution.

From N. Y. State Mus. Bul. 19, pl. 116

## Plate 34

Palisades-of-the-Hudson, as seen from Hastings, Westchester county. The rock is a sort of dark colored lava (diabase) which presents a crude columnar structure and rests upon sandstone of Triassic age. The sandstone is largely concealed by the rock debris just above the river level.

From N. Y. State Mus. Bul. 19, p. 4

Plate 35

Looking eastward down the valley of the west branch of the Susquehanna river from a point 2 miles west of Renovo, Clinton county, Pa. The river here cuts across a series of Appalachian ridges and the view was taken from near one of the summits. The remarkable concordance of altitudes or "even sky line" of the ridges is well shown and this uniform level of the ridge tops represents the surface of the upraised Cretacic peneplain.

Photo loaned by W. M. Gaylor, Sag Harbor, N. Y.

Plate 36

Precambric rock in the Highlands of the Hudson. Anthony's Nose and Manitou mountain along the Hudson river. The fairly even sky line of the mountain summits represents the old, upraised Cretacic peneplain.

From N. Y. State Mus. Bul. 19, pl. 8

Plate 37

Normal fault in bank of East Canada creek near Manheim, Herkimer county. The heavy beds on the left are Little Falls dolomite and the thin beds on the left are Trenton limestone. The steep tilt of the Trenton strata was caused by friction producing a sort of up-drag effect, during the slipping of the rock masses. The hole is artificial.

From N. Y. State Geol. Rep't 1894, pl. 4, facing p. 4

Plate 38

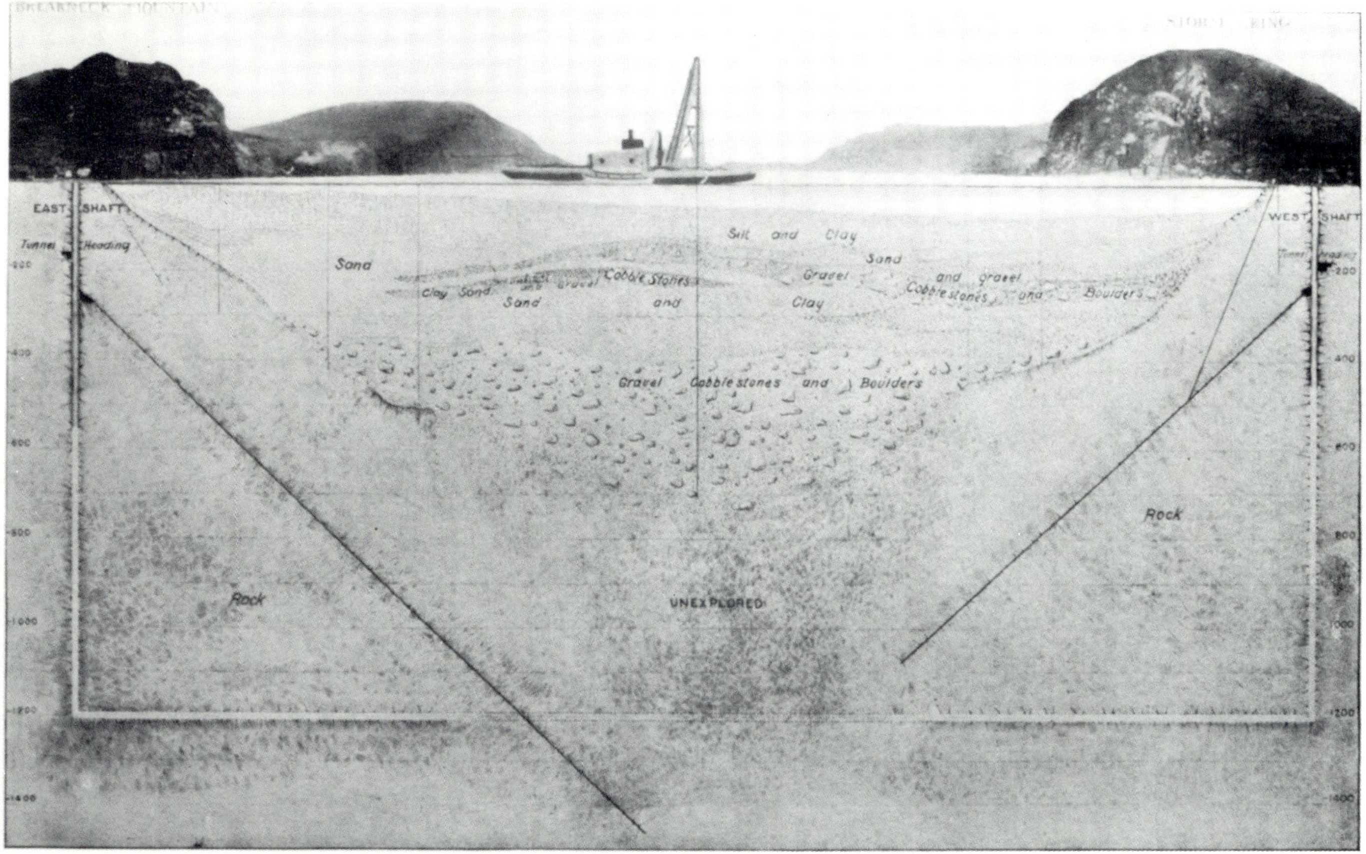

A composite diagrammatic cross-section of the Hudson gorge at the entrance to the Highlands-of-the-Hudson between Storm King and Breakneck mountains. The shaft and tunnel (at a depth of 1200 feet) show the position of the New York City aqueduct. The bottom of the old Hudson channel is seen to be deeply buried under clay, sand, gravel and boulders. The center boring has penetrated more than 700 feet without striking the old rock bottom. The depth of the rock channel here is more than 2000 feet or from the top of Storm King to the bottom of the buried channel. [illegible]

Plate 39

Glaciated surface of Onondaga (Devonic) limestone at Cheektowaga, Erie county. The passage of the great ice sheet over this ledge of rock wore it down to a generally smooth surface, except for the scratches which were produced by the abrasion of small rocks or pebbles at the bottom of the ice.

From N. Y. State Mus. Bul 19, pl. 94

Plate 40

A glacial boulder or "erratic" of Precambric syenite in the bed of Black river, 2 miles northeast of Boonville, Oneida county. The boulder is 27 feet across and 17 feet high and rests upon Black River limestone. It has been transported some miles at least.

Photo by W. J. Miller

## Plate 41

Drumlins in Wayne county, 3 miles north of Walworth. Looking east of south.

From N. Y. State Mus. Bul. 111, pl. 25

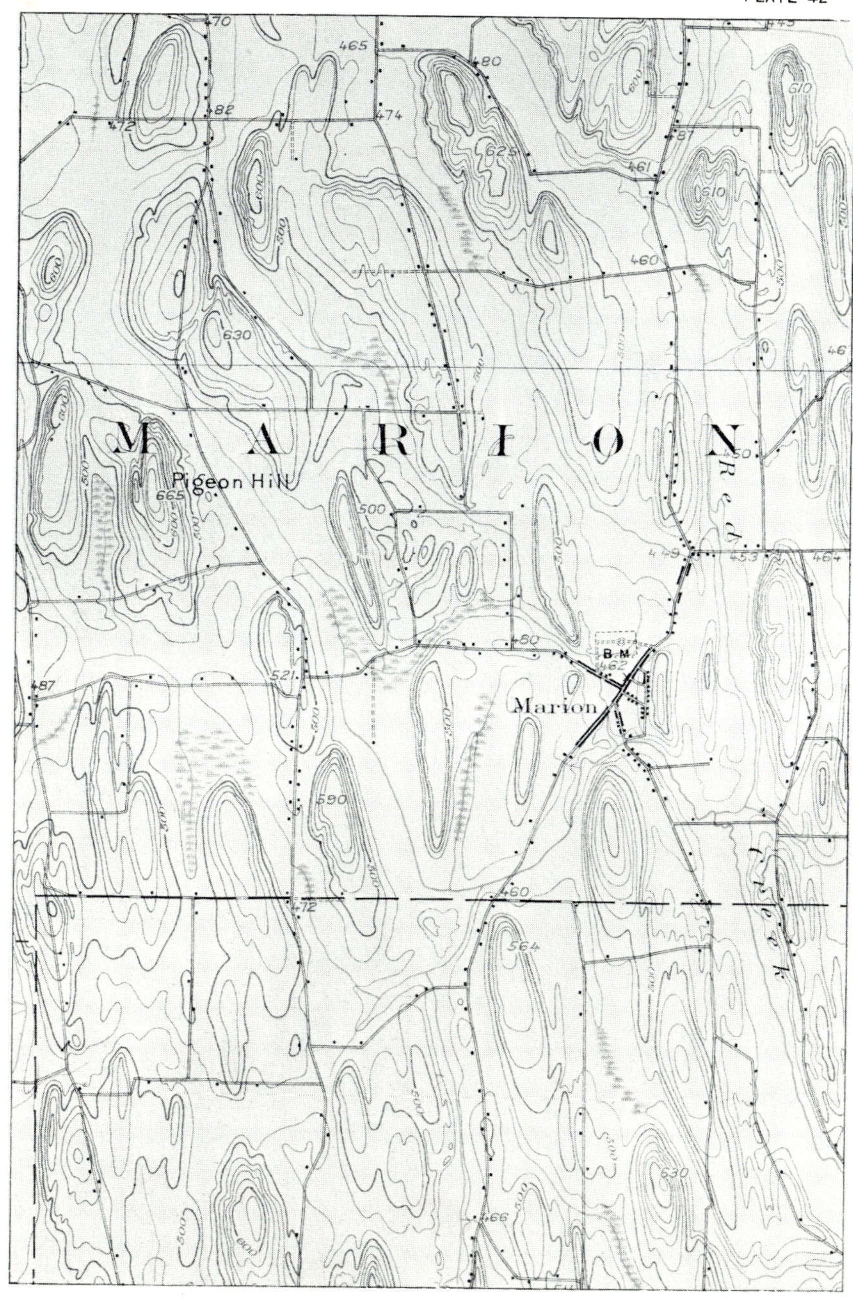


A topographic map illustrating that part of the Ontario plain which is studded with low, elliptical-shaped hills (drumlins) of glacial origin. Practically every hill seen from the train window between Syracuse and Rochester is a drumlin. From the Palmyra (U. S. G. S.) quadrangle. Scale, about 1 mile to the inch.

Plate 43

Typical glacial topography as seen near West Junius, Ontario county. The low, irregular hills all consist of stratified glacial debris and are known as "kames."

From Annual Rep't N. Y. State Geol., 1894, pl. 8, facing p. 80

Plate 44

General view of the gorge at Little Falls, looking up the Mohawk river from the eastern part of the city. In the bottom of the gorge is Precambric igneous rock (syenite), while the steep hills on either side of the river are of Little Falls (Cambric) dolomite.

Plate 45

High Falls in the gorge at Trenton Falls, Oneida county. Both the upper and lower portions of the falls are shown, the total descent of the water being 126 feet. The rock is Trenton limestone, at its classic locality, and it abounds in early Paleozoic fossil shells. The gorge here is 200 feet deep and wholly of postglacial origin.

Photo by F. B. Guth, Utica, N. Y.

Plate 46

A nearer view of the upper portion of the High Falls, at Trenton Falls, Oneida county. The height of this fall is 50 feet.

Photo by F. B. Guth, Utica, N. Y.

Plate 47

General view of Niagara Falls from the American side

From N. Y. State Mus. Bul. 45 pl. [illegible]1

Plate 48

The Horseshoe falls with Goat island and the Three Sisters, as seen from the Canadian side. This view illustrates the narrowness of the gorge now forming.

From N. Y. State Mus. Bul. 45, pl. 11

## Plate 49

The Whirlpool rapids and American bank of Niagara gorge, looking north. The rocks are nearly horizontal strata of Siluric age consisting of Lockport or Niagara limestone (forming uppermost cliff) followed downward by Rochester shale, Clinton limestone and shale (exposed near middle of bank), and Medina sandstone and shale. The gorge is here 250 feet deep. From N. Y. State Mus. Bul. 45, pl. 10

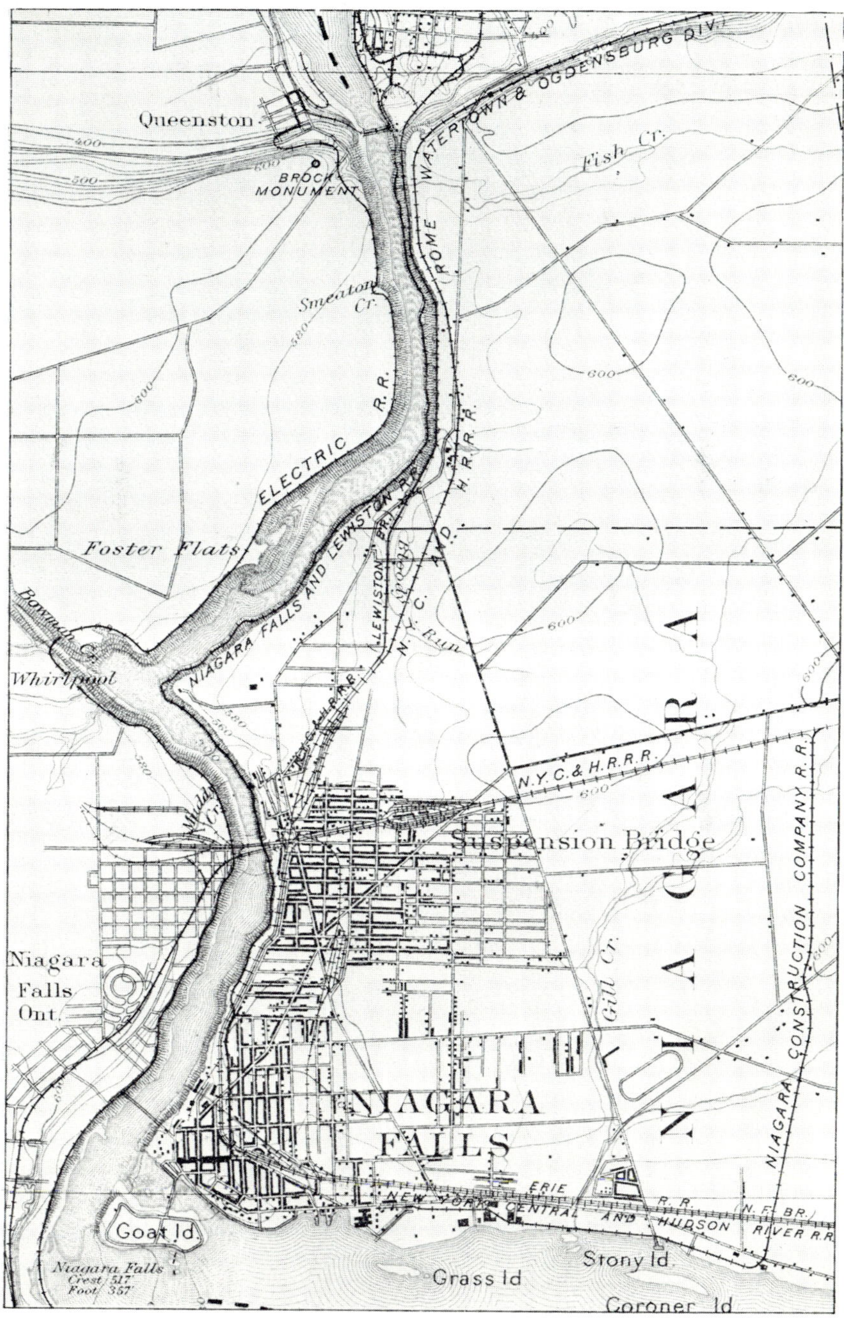


Portion of the Niagara Falls (U. S. G. S.) quadrangle, showing the position of the Falls, the long, narrow gorge cut through the very level Ontario plain, the position of the Whirlpool and the "Niagara escarpment" at Queenston. Scale, about 1 mile to the inch.

Plate 51

Middle falls and the gorge of the Genesee river just below Portageville. The falls are 110 feet high and the exposed rocks are shales and sandstones of Portage (Devonic) age. The gorge is postglacial in origin.

From N. Y. State Mus. Bul. 118, pl. 15

Plate 52

Triphammer falls and gorge at the edge of Cornell University campus, Ithaca, N. Y. The rocks are dark, thin bedded shales of Lower Portage (Devonic) age. The gorge is postglacial in origin.

From N. Y. State Mus. Bul. 19, pl. 81

AF615210